TRUCKPATCH:
A Farmer's Odyssey

By Ward Sinclair

The award-winning Washington Post columns
about the joys and sorrows of producing organic
food for body – and soul – at
Flickerville Mtn. Farm & Groundhog Ranch.

Edited by Cass Peterson and George DeVault

Published by

The ★★★★★★★★
**AMERICAN
BOTANIST**
booksellers

CHILLICOTHE, ILLINOIS

Truckpatch:
A Farmer's Odyssey

By Ward Sinclair

The award-winning Washington Post columns
about the joys and sorrows of producing organic
food for body – and soul – at
Flickerville Mtn. Farm & Groundhog Ranch.

Edited by Cass Peterson and George DeVault

First Edition

Cover painting: An original oil by Leila Finn
Cover design: Jane Colby
Inside text design: Ed Courrier
Copy editor: Melanie DeVault

Published by
American Botanist Booksellers
P.O. Box 532
Chillicothe, IL 61523
Phone: (309) 274-5254
E-mail: agbook@mtco.com

ISBN 0-929332-11-3

*To all farmers, everywhere,
and "the farmer" in us all.*

CONTENTS

——➤◦◅——

IV: FRIENDS AND NEIGHBORS

V: THE BUSINESS OF FARMING

PREFACE

—————➤•○•◄—————

This book is a collection of essays written by Ward Sinclair for The Washington Post from 1989 through 1992, after he had left his position as national correspondent to become a full-time farmer. The essays appeared on a sporadic schedule, roughly once a month, in The Post's Food section, under the name Truckpatch.

The essays tell about the simple yet complicated life of a small-scale farmer, a man who grows vegetables for people to eat. That's the simple part. No life is ever completely simple, of course, least of all a life so completely dominated by the benevolence or malevolence of nature.

There are frustrations and disasters in this simple farming life; there are joys and revelations as well. The farmer's topics range from the pleasure of watching a seedling break through the soil in the full promise of spring to the heartbreak of watching crops wither and die for lack of rain in the critical heart of the summer.

Ward draws well his characters, and they are not always human. Deer, chicken manure, black plastic mulch, nameless seed companies — these, too, are part of the cast. So are rural neighbors, friends, market customers, and the young people who toil with the farmer in the Truckpatch.

The columns would not have been written except for the persistence of Bob Kelleter, then The Post's food editor and a subscriber to the weekly bags of produce that Ward delivered to former colleagues.

A born schmoozer, Ward liked to regale his friends with tales from the truckpatch when he came into the office with his weekly veggie deliveries. Bob thought the reports ought to be written down and shared with The Post's readers.

Ward wasn't initially keen on the idea. Having cast his lot as a farmer, he professed to have no interest in ever again setting words to paper. But eventually Bob was persuasive. Ward agreed to write periodic dispatches with the proviso that the essays would be written in third person. Modesty would not permit the farmer to use the pronoun "I."

The essays got their share of critical acclaim. In 1990, Ward was named top food columnist by the Newspaper Food Editors and Writers Association. More important to him was that his little vignettes of farm life won loyal readers, many of whom wrote notes to the farmer after each column appeared.

When Bob Kelleter moved to Maine and a new editor took over The Post Food section, the essays stopped. It was time, Ward said, to move on to other things.

In 1995, three years after the last Truckpatch column, Ward died. Since then, his many fans have asked repeatedly that the essays be gathered into book form. I apologize that it has taken this many years to do that task, and I extend my thanks for their patience.

Special thanks go to George DeVault and Keith Crotz, who never nagged but always let me know that they were willing to help see Truckpatch into print. The same to Bob Kelleter, who had the vision to see the value of Ward's musings about agriculture, food, and the special bond between farmer and land.

Thanks, too, to the Miliuses, Peter and Polly, for their role in launching the small farm that became the inspiration for the essays, and to the Finns, Barney and Jane, whose friendship helped the farmer and his partner through many a rough spot.

A few editor's notes: the essays do not appear here chronologically, but have been re-ordered into broad categories. The largest of these categories is the last, titled The Business of Farming. Ward's readers found entertainment and inspiration in his words, but he had the ulterior motive of education as well.

Ward was not a gentleman farmer, and the Truckpatch is not Walden. It was and remains a working enterprise. He intended his essays to instruct his readers on the nitty-gritty details of small-scale farming, and hoped they would share with him the passionate belief that such farms have an important role in American agriculture.

The text is as it appeared in the pages of The Post, with a few minor exceptions. The essays were written on a typewriter, occasionally on a computer, and sometimes even by hand, and were dictated over the telephone to a typist at The Post. Because of that, there were occasional dropped words or misconstrued words, and these I have corrected to the best of my ability.

Cass Peterson
Flickerville Mountain Farm & Groundhog Ranch
January, 2000

PROLOGUE

⟶⟶◦◦◦⟵⟵

It wasn't our idea to buy a farm. It was Peter Milius' idea. No, it was probably Polly Milius' idea.

Peter Milius was (and is) an editor at The Washington Post. His wife, Polly, was (and is) a Justice Department lawyer. Their friendship with Ward went back to the '60s and The Louisville Courier-Journal, Ward's old newspaper. It may well have been in the interest of saving that friendship that Polly decided that Ward should own a farm.

The Miliuses owned a farm of their own – a couple hundred acres of woods and hilly pastureland – in Fulton County, Pennsylvania, two hours by interstate highway and another 30 minutes by narrow country roads from the nation's capital.

The place had an enormous, rambling farmhouse, a sizable barn, and a resident herd of heifers owned by a nearby dairy farmer who apparently had trained them to arrange themselves scenically according to the angle of the sun on the hillside.

The Milius family used the farm only as a retreat, to Ward's utter dismay.

Ward was a fervent gardener who chafed under the constraints of a suburban garden plot – all subsoil and clay, lacking in sun except in the front yard, where I insisted on planting flowers, and too small for melons or sweet corn.

The Milius farm, on the other hand, was a spacious piece of land – decent land, from the looks of it – with sun all day long and a copious supply of cow manure right there for the taking. To own such a piece of land, and to build upon it a tennis court (which they did) rather than to plant an apricot orchard or an asparagus patch, was to Ward's mind the basest form of human folly.

He wasn't the least bit shy about letting the Miliuses know this, and, frankly, Peter and Polly got tired of hearing it.

In July 1983, Ward and I were in Manchester, New Hampshire, as part of a team of reporters doing advance work on the 1984 presidential campaign, when Polly cannily figured out how to put Ward in his place – literally.

Peter made the telephone call. "There's a farm coming up for sale in Fulton County next weekend," he said. "And you're going to buy it."

It wasn't hard to sell Ward on the idea. For several years, he had served as The Post's agriculture correspondent. He'd tramped corn fields in Iowa, wheat fields in Montana, apple orchards in Washington state. He'd sat around dozens of kitchen tables in dozens of farmhouses listening to farmers talk about their work, their lives, and their dreams. For most of those farmers, the work was getting harder and the dreams were fading fast.

Take-it-or-leave-it prices, tight credit, rising production costs, global competition. As the per-acre profit of farming narrowed, the farmers Ward talked to tended to fall into two categories: those who were getting bigger, and those who were getting out.

To agricultural economists, small, family-operated farms were as old-fashioned as butter churns and just about as useful. Farms with limited acreage and small-scale equipment were pleasant additions to the countryside, maybe worth preserving for the sake of picture postcards, but hardly worth consideration as a force in agriculture.

Ward begged to differ. He believed firmly that there was still a place for the small operation in American agriculture, and it wasn't just his agrarian idealism talking. Ward liked to cook (and eat) as much as he liked to garden, and the poor quality and limited selection of fresh foodstuffs in the grocery stores irritated him no end.

Ward specialized in Mexican food – real Mexican food. He learned to cook in Mexico where he attended college and then worked at a newspaper in Mexico City. The cilantro he needed for his Mexican specialties was insipid, and the peppers were bland. The snap beans were flaccid, the eggplant wrinkled, and the California strawberries tasted of jet lag.

As frequently as we could, we shopped at roadside stands. Even then, the selection was sometimes disappointing or tired-looking, consisting of the same bland varieties found in the supermarket and often trucked in from the same produce terminal.

Ward was convinced that there was a market for top-quality, fresh produce. If that market could be found and satisfied, it might be the salvation for at least some of the farms being crowded out of the commodity markets.

Since the ag experts and land grant universities didn't seem interested in showing the way, Ward determined he'd just do it himself.

For my part, I had only two requests. First, if we were going to run a farm, it would have to be an organic one. My beat at The Post was environment and natural resources, and I'd covered enough toxic waste and pesticide-poisoning stories to want no part of a profession that relied heavily on chemical crutches.

The second was that this farm had jolly well better have plenty of space for flowers.

There was one slight drawback to Ward's plan. Neither of us had any real farming experience.

We were both of rural background. Ward was from Bloomington, Illinois, where his novelist father (Harold Sinclair, author of "The Horse Soldiers" and other titles) had kept a large garden and a goat. I was from western Kansas. My uncles ran wheat and cattle ranches, but my folks ran a small service station and café.

At our tract house in Alexandria, Virginia, our vegetable acreage consisted of three 4- by 8-foot plots in the side yard. The plots, visible from the street, were framed by landscape timbers so as not to offend the sensibilities of the neighbors.

We ate from the garden and made gifts from the garden, but we'd never sold so much as a sprig of parsley. Nor did we own any equipment beyond spades, rakes, and hoes. Not even a power tiller.

But on a strangely chilly day in August, 1983, we were part of a crowd gathered in the drought-dusty front yard of a shabby farmhouse in Fulton County, Pennsylvania, watching an auctioneer sell off the stray pieces of a lifetime of farming.

It was an estate sale. The farm, listed in the auction bill as 65 acres, had once housed chickens and pigs. The derelict coops and sties still stood, or partly stood. The owner, long retired and recently deceased, still kept a few cows, but those had been taken away.

It had been a dry year, and the land was parched. In an effort to spruce it up for the sale, someone had mowed the fields and baled the cuttings for hay. On some 30 acres of tillable land, there were perhaps 10 large round bales scattered over the landscape. The garden on the north side of the house held a dozen scrawny tomato plants and a few cabbages, cracked open and spoiled.

The farmhouse was two stories tall with a sagging front porch, and covered with ragged asphalt siding. A quick inspection determined it had no central heat, minimal plumbing, and two electrical outlets – one in the kitchen and one in the living room.

We bought it.

When the hammer came down, Ward looked over at Jack Strait, who had been standing next to him during the bidding. Jack ran a dairy farm near the Milius place, and city people seemed to amuse him.

"Jack, all I wanted was a place to grow a quarter-acre of carrots," Ward said.

Jack shook his head in complete disbelief.

The Flickerville Mountain Farm & Groundhog Ranch got its name over several glasses of cheap jug wine at the end of yet another long day of clearing brush, tilling sod, spreading manure, and repairing bad plaster.

To us, a quarter-acre of growing beds, plus a used Troy-Bilt tiller, a high-mileage used pickup truck, and a chain saw constituted a farm. Flickerville Mountain came from old maps that gave it as the name of the ridge to our west; the groundhog part was obvious the first time we strolled through our overgrown pastures.

The neighbors, who drove by slowly on weekends to watch us bent earnestly over our work, called it "the garden." It was many years before we accepted that term and learned to regard it as a badge of honor.

Despite the "farm" in its name, Flickerville Mountain Farm was a market garden, devoted to the hand cultivation of vegetables, fruits, herbs, and flowers. The gardens expanded slowly as we learned what we needed to know about growing and marketing produce.

A year after we bought the place, Peter Milius asked Jack Strait what he thought of the strange goings-on at the old farmstead.

"Well," Jack said, stroking his chin thoughtfully. "They're either going to do a whole lot more, or a whole lot less."

For five years we were part-time, commuter farmers. Through the good offices of Peter Milius (it does help to have your boss on your side), we were able to swing three-day weekends at the farm during the growing season in return for six-day weeks at The Post most of the winter.

We initially sold to an organic wholesaler, which served to cement Ward's belief that this type of farming would only work if there were no middlemen between farmer and eater.

We sold at farmers' markets, we sold to restaurants, and we sold to our colleagues at The Post through what we called a "subscription service." Later, we learned that the weekly veggie deliveries were a variant of a marketing idea from abroad called Community Supported Agriculture. Ours was among the first working examples in the United States.

It was much tougher than we had anticipated. Ward was determined that the farm be a working enterprise, not a place for weekend relaxation and puttering.

In the early years, friends came often, promising to help out on the farm but really expecting an antique-filled country house with a little garden, and relaxed hosts just in from milking the chickens or something.

What they found was two people in dirty dungarees, a house furnished at yard sales, self-serve beer in the refrigerator, and frozen pizza on the dinner menu. We made an informal rule: if you come to visit at lunch time, you bring it.

In 1988, we were attending two farmers' markets and delivering weekly to an organic wholesaler, as well as packing and delivering about 50 bags of vegetables every week to Post colleagues on our subscription list. This was done from Friday through Sunday. The rest of the week we rested up at our newspaper jobs.

At the end of that season, we were completely exhausted. But the farm broke even. Receipts from sales covered the mortgage, the equipment loans, the seeds, the tools, the fuel, the occasional help from local teenagers.

There wasn't a dime for either of us, of course, but it was promising enough for Ward.

"Let's do this full time," he said.

And so we did.

I.
THE FARMER
AND THE LAND

THE SWEET SMELL OF SUCCESS

Feb. 20, 1991

It had been a confining winter and when the first kind day of February arrived, the farmer avidly attached the plow to his tractor and headed to a field that had been bothering him for months. The ground should have been prepared last fall, since it was to hold part of the new year's potato crop, but other chores had kept the farmer from the field and it had nagged at him incessantly.

The soil had been thawed for some days now, the moisture had drained and the task could wait no longer. There still would be time to ready the plot for April planting.

When the sharp points of the plow cut into the turf and turned up curling slices of soil, a dense, musty aroma rose, and the farmer realized that this was what he had been missing all through the winter break – the indescribably comforting smell of good earth.

As the farmer slowly moved his machine in long sweeping furrows running counterclockwise, the fragrance grew and aroused the senses. By an instinct that mysteriously has come to him, the farmer knew from the smell that this field could be very good for his potatoes.

Now, there is no denying that the study of soils has become just as complex a science as the study of the human body, and the farmer pays due respect to the experts who have made that come to pass. He reads and he listens, and he is impressed by the imparted knowledge.

It takes no great wisdom or classroom teaching to understand that a healthy soil is a nation's most vital resource. Food abundance and the social well-being that derives from it are tied to the way a land is husbanded.

So as a duty, the farmer follows the rules as best he can. He wants this small piece of land to make its proper contribution. He wants to leave it healthier and more productive than he found it. And he gets satisfaction from knowing the source of his motivation.

The farmer takes his soil samples each year and sends them off to laboratories for analysis. When the findings are positive, covered with little smiley faces, he is elated. When something is amiss, he feels poorly and devotes himself to the recommendations for improving this field or that field.

But along the way, perhaps from spending so much of his time on bended knee with his hands immersed in the soil, the farmer's awe of the great knowledge of the scientists and technicians has become tempered. He has learned that cold laboratory findings cannot replace a tiller's instincts.

The laboratories and their tests reduce the farmer's land to charts and numbers. They report back in terms that the farmer does not fully understand, and the contradictions often leave him reeling. At times the farmer feels small and inadequate for his inability to parse the language of the science.

Other times, the farmer is lured by the commercial propagandists' claims of magic bullets and special formulations that promise the new Eden. He is tempted to allow himself to fall prey to quick solutions that could, for example, give him a wondrous abundance of potatoes.

In the end, however, the farmer relies more and more on his gut feeling – his instinct, as it were. It is as simple as knowing that he must put back in to the soil at least as much as his plants extract from it.

His choices are to succumb to the magic bullets, or to spend less money and laboriously increase his soil's wealth by using the plant and animal residues that nature has provided him. The instincts lead him to the latter course, harder though it may be, for this intimacy with each of his fields has shown how this works.

From the smell and the touch and the color, the farmer has come to sense if good things will happen in each swatch of ground. From the texture and the lay of slope, he has begun to know which crops will work best in each of the patches. From these same nuances, he is coming to know if – and where – he may have certain pest problems.

So while the farmer often feels inferior for his lack of formal knowledge and for his inability to talk the language of the experts, he gets a certain quiet comfort from the teaching of the soil through his hands and his nose.

There are friends and neighbors who make some small light of the farmer's obsession with being out there on his tractor. That is all right. Each trip to the field, the farmer reasons, is akin to boarding a school bus that will take him to a temple of learning.

This soil, after all, is a temple of sorts. The learning is constant, the feeling is comforting, and the smell ... the smell imparts a rare and mystical incense.

MIGHTY HOPE FROM LITTLE SEEDLINGS GROWS

Feb. 22, 1989

The farmer is much like the tree. As the days turn longer in the waning winter, the time of dormancy begins to ebb and the sap rises in anticipation of resurgence and renewal. No formality announces this beginning. It is a feeling in the bones, if you will, that culminates in a rush to gently drop the first seed into the soil.

The goal is to produce a perfect vegetable. But no matter how exquisite it may look on a truck farmer's stand in June, the spear of broccoli, say, will not begin to reflect the intense collaboration between the grower and seed that began in February.

And so it is with tomatoes, melons, squash, and other vegetables started subsequently in the greenhouse and given much the same personal attention. The basis of their perfection rests on this.

The feeling of rising sap, alas, rarely coincides with the dictates of nature. So here in the mountain climes, where winter will remain cantankerous and unpredictable for many more weeks, the farmer must outwit and assist nature.

His coconspirator is the greenhouse, a plastic-covered structure that traps the winter sun's warmth to simulate the balm of spring and allows the sap of man and plant to rise apace. Seeds started here will become small plants that will go outdoors when winter breaks.

Not to read too much into this, but the plant's triumph carries more than economic importance to the farmer. It represents, perhaps above all, a reaffirmation of self and nature that man seems to require. The plant's progress will renew or dilute his wonder for life. His awe of mystic processes will be re-consecrated; his sense of personal creativity and competence lifted or deflated.

At every step of the way, the farmer and his plants will maintain intimacy. They will speak to each other, become friends. Each will understand the other's idiosyncrasies. The farmer will be a protector and nurturer, a friend who will sunder attacking pests and assure a balanced diet for the plant.

To some degree, this is the compelling spirit on every farm. But here the difference between large farms and small farms becomes clearer. It is basically a difference of intimacy – with the seed, with the soil, with the individual plant, with nature. The difference between factory-made and hand-built applies to vegetables as well as autos.

Take, for example, the broccoli, with which a relationship is being established here this February. More than 4,000 broccoli seeds, each barely large enough to feel between thumb and finger, were planted one by one in the greenhouse flats early this month.

This ritual, introducing seed to soil, flames the anticipation. After the seeding the farmer would observe his trays several times each day, awaiting the reaffirmation. The first signs were tiny fissures on the surface of the soil in the flats. Then came the first plants – minuscule, dark green, round-leafed much like a clover.

Within weeks, properly warmed and watered, these plants will acquire strength and size until they are ready to be sent into the fields to continue their new life. They will be cozied just as a new child in the family would be flattered and protected.

Each small plant will be gentled by hand into long beds covered with black plastic, which warms the soil, retains moisture, and discourages competition from weeds. Each seedling will receive a drink of water and a taste of fertilizer that will help it fight transplant shock and speed early growth.

Then gauze-like tunnels that allow rain and sunlight to reach the plants will be erected over the beds to protect against the cold. When true spring is at hand, with the threat of hard freezes past, the tunnels will be removed so the plants can grow unfettered.

All the while, the farmer will regularly patrol for insects. If they appear, which they almost surely will (another reaffirmation of life), they will be dispatched with organic sprays that are toxic only to the bugs.

The farmer's patrols will also disclose the nutritional state of the plants. If it is indicated, a jolt of fertilizer will be applied. Just as the plant begins to form the button that will become the head of broccoli, it will be given another dose of food – a bit of sea kelp, liquefied fish waste, or a tea derived from animal manure.

Yet at best it is all a tenuous process. Nature and the weather it provides will be the final arbiters here. Last year, for example, even as all the requisite care had been taken and attention given, a sudden hailstorm shredded hundreds of flourishing plants. Many died and many others recovered too late to make a crop.

This year, the farmer has decreed, there will be no hail. He has decreed that spring will provide only the warm-but-not-hot days and cool nights that are essential to his broccoli. He has no special influence with nature, but he is confident that spring rains will be ample, if not abundant.

He is proud, or foolhardy, enough to imagine that his broccoli will be better than other broccolis, and certainly better than the broccolis grown on an assembly-line farm in some parts of the country where the grass always seems greener and the elements more congenial.

Yet the farmer is also realist enough to know that if his commands are ignored, he will continue to try. Just as his spring broccoli is coming to harvest size — assuming it does — he will be sowing in flats the seeds that will become his fall broccoli crop. And if that fails, he knows there is always another year. It is as certain as the humble broccoli seed's reassertion of itself each February. Nature, treated with respect, offers more than one chance.

Sowing the Seeds of Hope

Dec. 26, 1990

A visitor came recently from California, bearing gifts of wine from a vintner friend of the farmer's, and the challenge was to find a proper response. The farmer could not offer an "estate reserve" or some other vineyard delicacy, so he turned instead to a small packet of seeds that he had carefully saved during the summer for replanting in 1991.

This seemed the best and most meaningful response imaginable.

These were the seeds for an expensive European sweet pepper that the farmer feels very possessive about. He selects out the finest of these peppers each year, saves the seeds, and holds them closely lest his competitors get any ideas.

As the farmer drank his gift wine (a superb cuvée, it turned out), he thought about his growing obsession with seeds of all kinds and how they have come to dominate much of what he does in the truckpatch.

The seeds, after all, are the source of life for his plants, and while the farmer does not begin to understand the deep mystery of their power, he views them with increasing respect and even reverence.

Spring, summer, or fall, the farmer never fails to clutch with emotion at the sight of first emergence of his newly planted seeds. The thrill is so intense that the farmer often squanders precious time just to visit his most recent plantings to check for emergence. He will drop to hands and knees, looking intently for the first sprouts.

The success of planting has to do with the financial security of the truckpatch, obviously, but it is more. The emergence of those first seedlings is an important sort of affirmation that the farmer has done something right.

There is a tendency to take all of this very personally and when a crop fails, the farmer is usually wont to become despondent and to blame his own ineptitude for the refusal of his seeds to germinate.

Last spring, for example, when it seemed the weather was ideal, he put in the first of his green bean plantings. It was warm enough and the rain was abundant at the right time, but only one out of many thousands of the bean seeds germinated.

8

He blamed himself instantly, surmising that he had been too early or that the ground had been too wet. As a precaution, though, he used no more of that particular expensive batch of seeds because he feared another failure.

There was no early crop, but nature was generous. Later plantings of other bean varieties did better than ever before. They were so productive, in fact, that when a fall labor shortage developed, the farmer was forced to simply abandon the patch because of lack of time.

It was a painful act, seemingly insensitive to those seeds' desire to be of use and the farmer felt badly about it.

This winter, contemplating another early spring bean planting, the farmer announced that he would give those leftover seeds one more chance. His partner wisely insisted that they be put through germination tests before plans were locked in place.

Virtually none of the beans germinated and so, with heavy heart, the farmer relegated the seeds to his compost heap along with the seeds of peas and corn that also flunked the test.

If the seeds hold sway over his fortunes in the warm months, it is no less true in the off-season. Much of the farmers' idle time in the cold months is devoted to devising planting schemes for the new year, studying old and new information about varieties and deciding which of these would work in this part of the world.

This is the time of year when literally dozens of seed catalogues fall into the farmer's hands, each seeming to promise a more bounteous crop than the other and each trumpeting the glories of all the new hybrid varieties that seed scientists spend years perfecting.

The farmer himself spends much time weighing the supposed benefits of all these new products, but he knows also that seeds are huckstered no less vigorously than laundry detergents or new cars. And he must be acutely aware of costs – the same varieties can have wildly differing prices for no apparent reason. So he must proceed with caution, for he has come to understand that last year's seed marketing sensation may turn out to be this year's flop. There is nothing like the collective experience of farmers and gardeners to tell the breeders where they've gone astray.

One of the farmer's suppliers admitted ruefully in a recent mailing that the new hot pepper he had worked so long to develop and that he had ballyhooed relentlessly had been a marketing flop. It was gone from the 1991 catalogue but, perhaps to compensate, the breeder brought back by popular demand a Japanese squash that earlier was banished from the warehouse.

And so it goes with dozens of varieties, some seemingly developed with no thought of practicality. They are here today, gone tomorrow. Just when the farmer becomes comfortable with a new sweet corn or pepper or tomato, it often will vanish from sight.

Some time ago, after much searching, the farmer found the "right" sweet corn, a variety called Wonderful. Here, finally, was a corn that was worth growing. But it has disappeared from the catalogues, despite the farmer's assiduous search for it each winter.

He will try many of the new ones and he will drop some others that did not meet his standards during earlier trials. He has still others on his just-one-more-chance list for 1991. And he will retain those that performed well and that sold well at market in years past.

His best friends, though, will be those non-hybrid varieties that he has been able to save and grow with success. Not estate reserve, in the vintner friend's language, but gifts just the same.

Nature Changes Her Clothes

Sept. 13, 1989

S ome very nice-looking late tomatoes are scaling the trellises in the main field and if all goes well they'll be red-ripe and marketable before first frost, but frankly the farmer doesn't much care. This happens with the onset of every autumn, although it is not to say that the farmer has lost interest.

The tomato, after all, has been a centerpiece of his anticipation and emotional investment since March and its contribution is appreciated.

But in a variety of subtle ways a change of seasons has been announced and nature is telling the farmer the time of the tomato has passed. The rhythms of growth go on and the farmer must look to the new things that rise in his fields.

The signs of change abound in the truckpatch.

The common fly has given way to the yellow jacket, a corsair in search of the sweet bounty of early fall. The cherry tomatoes ripen with urgency, knowing that their end is near. The zucchini plants give one last gallant flush of fruit. Days ago the fireflies and the hummingbirds stole away as quietly as they arrived. The night sounds are different now and the morning air brings a comforting chill.

All of this says that the tomato, coveted as it may be, must step aside for the crops that thrive in these latitudes in September and October – the months when the farm seems to explode in a final frenzy of lushness and plenitude.

The pumpkins are turning orange and the butternut squash shows the first tanning of maturity. Turnips already are bigger than golf balls. And with the cooler nights, the mustard and the kale have begun to crinkle and grow as they should. The rutabagas, fully aware of nature's announcement of change, bloat in size despite the absence of the usually predictable rains of early fall.

Small heads are forming in the broccoli and the purple cauliflower patches. Chinese cabbage and bok choy boom up in their beds. Lettuce, radicchio, fennel, and arugula, all faithful performers in the shorter and cooler days, seem pleased to be here.

11

It is as much a time of excitement as a time of regret. The excitement derives from the bounteous surge in the fields; the regret from the knowledge that another year's growing cycle, with all its squandered or mismanaged opportunities, soon will end.

The farmer will confess to poor judgments – not treating his tomatoes for early blight, underestimating eggplant demand, mis-timing the cutting of his spring lettuce – but he also would plead that nature dealt another fickle hand.

The spring was too wet and too cool. June and July were too wet and too cool. August, for all the promising thunderheads, was too dry (less than an inch of rain for the month) and too cloudy to be of much help to the heat-loving tomatoes and melons and squash.

There were more strange bugs than ever before. The moisture made the weeds and grass grow with abandon, brought fungal disease to the peas and peppers, and kept the farmer's machines from the field.

A sort of final comeuppance was delivered the other evening by six Holstein cows the size of steam engines, which found their way into a remote field of melons, pumpkins, and rutabagas after a fallen tree broke the fence that had kept them in a neighboring meadow.

The first alarm was sounded by a neighbor, who raced his red pickup truck out toward where the farmer was trying to figure how to deal with the dawdling fall crop of sugar snap peas.

The peas didn't look good – it seems they never do in the fall – but they were quickly forgotten. "Hey," the neighbor shouted, "there's a whole herd of Holsteins roaming around in your pumpkin patch." The neighbor then rushed off to give the cows chase.

Visions of a field of trampled and crushed pumpkins rushed to the farmer's mind. And maybe the watermelons would be ruined. And perhaps the cows would defoliate the rutabagas that had taken hours to nurture and keep free of weeds.

This had all the makings of a disaster, and in a year of complete unpredictability and uncooperative weather it was another blow that the farmer neither needed nor deserved.

The owner of the fence and the cows was alerted by telephone, and the farmer then wheeled his truck down toward the pumpkin patch to survey the damage. He silently thanked the helpful neighbor and pondered the wisdom of another neighbor who once said that a bad fence could ruin a friendship quicker than anything else in the country.

The cows were friendly but not inclined to leave their new grazing paradise peaceably. They were indifferent to the farmer's threats. They retreated slightly into a woods, then commenced an end run back toward the rutabagas.

The Holsteins' owner, a somewhat chagrined dairy farmer, finally arrived and herded the beasts back through the woods and into their own meadow. The damage was slight – no pumpkins or melons destroyed, maybe six feet of rutabaga tops munched off.

For his alertness and help, the neighbor was rewarded with a fresh-cut watermelon. To allay his chagrin, two melons were left on the front seat of the dairyman's pickup truck. The farmer felt only relief.

It seemed a good way to begin the new season of bounty.

ALL THAT IS GOLD
DOES NOT GLITTER

March 29, 1989

O n a flat stretch back by the woods, still high and wide enough to be mistaken from afar as some sort of pre-Columbian burial mound, sits a pile of farmer's gold. Actually, it's chicken manure, the remainder of a 23-ton load brought here in a semi-trailer months ago from Lancaster County, one of our country's richest farming areas.

Now, the shipping of chicken litter more than 100 miles has a certain social value that goes beyond just growing a tastier carrot. The manure will enrich the farmer's soil, it will help reduce Lancaster County's surplus waste, and, ultimately, have some small impact on the well-being of the ailing Chesapeake Bay.

Farmers in these parts, where dairy, beef, and poultry are major enterprises, typically dispose of their manure by spreading it on their fields to bolster soil fertility. But too much of it is a bad thing. The manure's nitrates leach into the groundwater, thence into the streams, and finally down to the bay, where they join other urban and industrial pollutants to stifle aquatic life and vitality.

Pennsylvania and other states in the Chesapeake watershed have put farmers on notice that they must become better manure managers to prevent contamination of the region's waterways with bacteria and nitrates. And some farmers, predictably, think they are being picked on.

In a way, they are right, for the manure problem really stems from our collective appetite. Anyone who eats meat and eggs and uses dairy products – from Lancaster or anyplace else – aids and abets the problem. City and country are in this thing together.

The trick, then, is to find ways to dispose of the manure that our eating habits generate without damaging the environment, yet allow the farmer to continue to sate the urban appetite for milkshakes and bacon cheeseburgers.

A happy side to the story is that many farmers, concerned about high costs and water pollution caused by commercial fertilizers, are beginning to learn a lesson that their grandfathers took for granted. Their discovery is that manure is a wonderful thing.

The companies that yearly sell farmers more than $7 billion worth of chemical fertilizers are not accepting this discovery with grace.

Their chief Washington lobbyist, Gary Myers, was apoplectic when the Agriculture Department recently dared suggest that farmers had been oversold on the benefits of the commercial fertilizers.

In another era, Earl Butz, a secretary of agriculture who loved to puff businesses like fertilizer manufacturers, used to joke that the country would go hungry if it relied on animal manures to fertilize the crops. Butz was incorrect. In fact, there is more manure around than can be reasonably imagined.

Consider, for example, that the typical dairy cow produces about 25,000 pounds of manure per year. Then consider that the country's dairy cow population is in the 10 million range. That's a massive amount of manure – and it doesn't even start to count the other huge amounts that come from broilers and laying hens, hogs and beef cattle, horses, goats, and sheep.

U.S. farm animals produce well over 1.5 billion tons of manure annually. That is enough to apply about 4 tons each year to every acre of land under cultivation. That is also sufficient to feed most soil handsomely. The problem is getting the stuff to where it is needed.

Thus, the farmer in the truckpatch felt a certain jubilation when the big green manure truck rolled up to his place early in the winter. His search to find the right stuff had been lengthy and it came at considerable expense, for the trip from Lancaster is long.

A driver named Earl threw a hydraulic switch and the truck disgorged a mountain of manure. The farmer circled the pile, which stood twice his height, and regarded it as a Christmas gift arrived early. He probed the pile with a shovel, even smiled as it emitted a gush of steam and the acrid stench of ammonia.

He got added pleasure from the thought that passers-by, especially other farmers, would be unable to ignore the pile. They could only surmise that more serious horticulture was about to take place on this curious little farm. In a country where dairy cows are the main source of manure, this richer chicken litter would become a sort of badge of purposefulness.

Beyond that flight of vanity, the farmer was excited about the benefit that the manure would bring to his crops in the new year. Nitrogen and minerals would bolster the plants and help improve the health of the soil.

As weather permitted, the farmer chipped away at his pile with a large shovel. He loaded the stuff into an old spreader, then hauled it out to the vegetable beds, where it had to be cast on the soil by hand because the machine could not spread it thinly and evenly enough. Then the manure was tilled into the ground to decompose and begin its enriching process.

There were days when the arms ached from the shoveling and lifting and there were evenings when the back complained. There were days when the cold and the wind seemed intolerable. Yet the prospect of a bountiful crop and a richer soil buoyed the farmer.

He knew that his 23 tons of chicken litter, removed from a place where they had too much, really was not even a flyspeck on the big picture. But it was a start.

And if it helped the Chesapeake become a place where watermen could again harvest their shellfish and if it would enable a young boy to someday learn the thrill of pulling in a rockfish, then the farmer knew his manure was better than gold. About that, he felt good.

GIVING FROST THE COLD SHOULDER

Oct. 30, 1991

The first frost came earlier than usual, toward the end of September, and it seemed a beneficent stroke of nature aimed perhaps at ending the farmer's summer of discontent over the weather. The heat and the lack of rain had made this the most difficult of years in the truckpatch, a year that the farmer secretly wished would end so he could get on with his optimistic planning for a new season.

First frost, of course, is always a definitive signal of seasonal change and an announcement of new rhythms and new colors on the farm. The farmer and his plants are inured to this change and ordinarily accept it without much complaint.

But this time something was different. The farmer accepted the quirk of an early frost, but his plants did not and it gave rise to pondering the magnificent, unfathomable forces of nature that regulate life and death in the plant world.

Many of these plants that usually crumple and blacken after the first dash of below-freezing weather declared that they were not ready to succumb. Their work was not finished – seeds had not been set – and it seemed they were determined to carry on until the mission was done.

The squash plants collapsed but they magically continued to set fruit and grow. The zinnias valiantly flashed new blossoms over the darkened foliage. The peppers, the raspberries, and the basil, always hypersensitive to frost, trembled a little but continued to produce.

This phenomenon continued through the second, the third, the fourth frosts, plants oblivious to the colorful change going on all around them. Brilliant reds, golds, umbers swept along the ridge and through the woods, a sure sign of fall, yet these plants kept on.

There are lessons here, although the farmer is not quite sure what they mean. If nothing else, it could be read as a message of hope and encouragement, something that every farmer needs from time to time.

For if these plants could summon the energy and the valor to go on about their business, despite their buffeting by the weather, why should the farmer do any less?

And how could the self-pitying farmer pray for an end to the season when his charges were telling him by example to hold his head high and to continue the struggle? Was it a subtle communique from Plant Central, urging the farmer to better appreciate the unseen forces around him?

Possibly so, for there were other points of inspiration all about the truckpatch where the farmer's plants were resisting the finality of fall, retaining their green, and fighting to produce.

Despite the lack of rain, which had retarded them badly, the broccoli and the cauliflower struggled to become full and set their heads. The kale continued to put on new growth, ignoring the dryness. The chard ignored the frosts and pushed up more dark green shoots.

From these events, the farmer takes succor. Wherever he looked, the farmer could see the inevitable signs of a season ending – the fall folk festival had come and gone, the fields were crisp with morning white, tired workers were drifting away, the north wind was turning sharper.

And yet so many of these plants that the farmer had nurtured and babied through the long and dispiriting summer were resisting, even rejecting, the signals and the admonitions of the weather. It was as though they had something important to say.

Now, there will be skeptics who make light, but the farmer has come to believe that these nuances are real and that there is some form of communication, albeit unspoken, between man and his plants.

In the truckpatch, the farmer from time to time actually has caught himself speaking kindly to his plants, occasionally giving a special one a name, or smiling benignly at his handiwork. It is not a cause of embarrassment to admit this.

It is only to say that communication of a sort goes on and from this the farmer can draw some hope. The plants are talking back now and they are suggesting, through the force of their struggle, that there should be cause for optimism in the new year.

That's the only way to read it, the farmer thinks, for otherwise he'd be left talking darkly to himself. Which isn't a happy thought.

STRANGE HAPPENINGS, WEATHER OR NOT

Aug. 8, 1991

It had been a trying day and when it was over the farmer went out to a field at dusk, lay down on the sun-crisped grass and contemplated a spectacular sky of pink buttermilk clouds. It was one of those perfect, precious times that occurs often on the farm – a moment of ultimate serenity, when the light and the shadows are just right and nature seems to reach out to enfold the farmer in reassuring embrace.

This is a magical security blanket that comforts the farmer and makes all his mundane problems small. It ameliorates the devilment of erratic weather, machinery breakdowns, intrusions by varmint and bug, and untold other difficulties.

As he lay there and gorged himself on that pink buttermilk, the farmer was moved to a small chuckle. Back at the house, the official weather radio was direly blithering about a tornado alert for the area. Of course, there was no tornado. Not even a rise in the wind.

How little we really know and understand, the farmer mused, about what is happening around us. For all the talk of ozone depletion and greenhouse effect and weather change and expansion of the deserts, all that we don't know about these things would fill volumes.

The farmer knows that strange things are happening to his climate, even though the weather data in his carefully kept notebook don't really confirm this.

When an unusual thing called a fig beetle showed up this year, apparently far from its customary habitat, the farmer knew something was amiss. When a water moccasin showed up in the farm pond of his friend Musachio, another creature way away from its habitat, he knew that something more was amiss. This is strange stuff.

Not long ago, the farmer met an entomologist, a man schooled in the techniques of chemical bug control, who expressed puzzlement about some of the non-toxic approaches being used to prevent insect damage in the truckpatch.

The scientist opined that these approaches would have to be studied to determine whether they really work. But, the farmer interrupted, they do work — we know that by experience. Well, then, said the scientist, we'll have to find out why they work.

Which is all well and good, the farmer thought, but when a crop faces a serious threat there is no time to find out why. So here in the country, given the uncertainty of all the expert advice that comes his way, the farmer relies more and more on his own findings and is guided more and more by his own hunches.

The early crop of Swiss chard was suddenly afflicted by dreadful looking black pock marks on the stems. The marks seemed not to affect the leaves, but they made the chard not very appealing.

The farmer could not decipher this problem, nor could his chard-growing friends, so he sent the leaves to be examined by a team of real experts.

No one had an answer for the pock marks, but one tentatively surmised that the problem might be a lack of boron. A farming colleague named Ricci, a man who knows his chard, snickered when he heard this conclusion. "They always blame it on a lack of boron when they really don't know what the problem is," Ricci said.

Now, none of this is to put down the sages and the scientists.

But in the absence of real answers to their problems, the farmer and his friends move in the same direction of following experience and hunches.

In the case of the chard, the farmer did what his intuition told him to do. He left it alone. And behold — the pock marks disappeared from the stems and the crop righted itself.

Another time, the farmer's sweet peppers developed a sort of rot that defied explanation. A university specialist pored over a reference book, diagnosed the problem, and recommended a fungicide so powerful and toxic that the farmer would not even consider using it.

Your problem, said another farmer who had traveled this same road, is not a fungus. Spray the peppers with Epsom salts and they will flourish. Ah, so, said the farmer, and he applied the spray. In a soil shown by tests to be magnesium deficient, Epsom salts — magnesium sulfate — was the perfectly obvious solution. The peppers flourished.

The farmer cannot wait for the definitive studies, but he concedes that in the long run the entomologist is right. Once we know the why, we may be able to better deal with the what and the how.

He would like to believe his occasional successes derive from his skill rather than his dumb luck, but the farmer knows better. He is perplexed and driven by the whys of a thoroughly perplexing 1991, a year of enigmatic weather and insect curiosities.

Why, for example, are there squash bugs on potatoes and tomatoes and other crops this year, but not on the squash?

Why did the Colorado potato beetle, which came so intensely this year that a statewide emergency was declared, suddenly vanish from the truckpatch? Where did it go and why?

Or why did the voracious Japanese beetle, appearing here in ever increasing swarms, just as suddenly disappear? Peterson, the farmer's cohort, as usual had many brilliant theories, which may or may not be correct. Yet these were only theories.

The farmer's inclination is to congratulate himself for another series of uncanny moves in pest prevention and damage control. He could do that, but it would be wrong. So until there are solid answers to why this or that occurs, he will continue to follow the instincts provided by experience and reason.

Returning to that pink buttermilk sky, on the evening of tornado alerts, the farmer wondered how the experts could get it so wrong. He was reassured to later read, after an unexpected storm blitzed the nation's capital, that the chief weatherman frankly admitted there's a lot about meteorology that even the experts don't know.

The weatherman actually could have been talking about the truckpatch. A lot goes on that the farmer doesn't understand. He knows more and more about what works and what doesn't, but he's never quite sure why. But he also has learned he can't wait for an expert to explain it.

FATAL DISTRACTIONS

April 22, 1992

Assuming himself to be the complete and thoughtful giver, the farmer years ago presented his partner with one of those little tape players with earphones, the better she should enjoy her weeding. Well, the tape player made it down part of one row and Peterson, without explanation or even apology, sent the wondrous little device to early retirement in a closet.

The farmer, hurt and chagrined, rescued the machine and took it with him to the fields to provide music while he rode the tractor. It was a fatal distraction, not unlike attempting calculus homework with "The Simpsons" blaring in the background.

Try as he might, the farmer could not concentrate. The tractor swerved erratically about the farm. No longer could the farmer hear the diesel engine, whose varied shifts of hum send vital signals to the operator. The sound of tiller strain and clank, still another arcane language, was stifled by the music.

So the farmer began to understand and Peterson's judgment was confirmed.

The tape player went back to the closet, but not before it had taught the farmer an important lesson: One can't improve on the natural symphony of the truckpatch, be it the babble of a mockingbird or the chunka-chunka popping sound of an old tractor.

The farmer has come to treasure what passes for bucolic serenity in these parts and it is for that reason that he swears at the frequent intrusions and tries mightily to resist the temptations that modern technology thrusts before him.

There was a time when the farmer, understanding that the truckpatch was a business, decided he must maintain contact with the world. He bought a portable telephone to accompany him in the greenhouse. It rang often and loudly, frequently disrupting the work and concentration with the banalities of market surveys, wrong numbers, and dippies.

Then a child broke the telephone's antenna and the farmer was rescued. Calls could not come in. The beguiling sound of silence came home again and the telephone joined the tape player in the closet. The twitter of bluebirds could be heard again, raindrops pounding on the barn's tin roof became special again.

The farmer admits to his eccentricities, but he still has trouble understanding why his neighbors insist on disrupting this serenity. One has a small television set mounted on the fender of a tractor. Another listens to a built-in radio while he plows. Another keeps a CB radio chattering constantly in his pickup truck.

And why, the farmer wonders, must his city-raised young helpers have stereo music blaring at the highest decibel levels when they are toiling hundreds of yards away? That insult was redressed when they blew a fuse in the receiver and were forced to work without a din.

There are still other intrusions on the silence that the farmer cannot control, but some of them provide information about life beyond the edges of the truckpatch. Many of these sounds have signatures that tell the farmer who's going where – that's Tanner's pickup, here comes Booth in his Jeep, young Carnell's riding that dirt bike again, sounds like Smitty's log truck heading to the mill, Bradshaw's tractor on another run to pick up hay.

There's a rumble far down the road, the unmistakable rumble of the school bus, and the farmer knows it's about 7:35 a.m. It must be mid-morning when the Pittsburgh-to-Hagerstown commuter plane buzzes overhead. The farmer knows it's just past noon when he hears the distant muffler of the mailman's car. It's 3:20 when the school bus rumble is heard again; 4:25 when a barely visible passenger jet flies to the west; 7:10 when Tanner's wife is headed home from work.

The farmer knows it's summer when the roar of what Peterson terms the "Fulton County mating call" jangles the ears. It's the muffler blast of a local blade's hot rod, announcing to female workers in the truckpatch that he's out, about, and available. The workers snicker, the farmer cusses.

The farmer has come to accept these intrusions, though begrudgingly, for there's nothing he can do to prevent them. But in the interest of keeping the peace he also has come to realize that he must just say no to the technology that he can control.

So no, a thousand times no, he will not buy that fax machine or the copier that farmer friends keep insisting are so key to keeping a truckpatch on the right course. He won't add another night light for security purposes, he won't replace the fuse in the workers' stereo.

And he will continue to leave a log uncut before cranking up the chain saw that can be heard a mile away. He will spade a small plot by hand before using the little tilling machine with the roaring engine. He will not turn on a radio even if it's time for the news. He will not rehabilitate the portable telephone, and the tape player will stay on the shelf.

And in a final declaration of independence, with quiet resolve, the farmer will not turn up the volume on his telephone answering machine. Enough is enough.

ROMANCING THE FARM

Nov. 22, 1992

It was a thoroughly miserable job on a thoroughly miserable fall day and the farmer, crawling on all fours in the dirt, thought he was entitled to ask why he had forsaken the warm comfort of the city for this kind of life.

The job was one that must be done every fall when the crops are in – removing the long strips of black plastic mulch that warms the soil, retains moisture, and stifles the weeds. The truckpatch seems unable to operate without this plastic, yet it is the bane of the farm.

The edges of the plastic, embedded in the dirt, shatter and tear. Muscles strain to break it loose. Dust or mud flies in all directions. The farmer's knees get bruised, his shoulders and arms ache, he is a human dirt ball. He wonders: Why bother? Who needs this grief?

But by the time the plastic removal is finished, Thanksgiving has drawn nigh – meaning another year's cycle of planting, nurturing, and harvesting is complete – and the farmer can consider his lot with a better sense of perspective.

So inured has the farmer become to the challenge of making each year a success that he cannot conceive of himself ever again sitting behind a desk, being paid well to do a job that had simply become too easy.

Here in the truckpatch there is no desk, the pay isn't great, and the job is never easy. But it offers something vital to every man – the freedom to succeed or fail on one's own. There is no one else to blame when things go wrong, no place to hide.

Yet over time, as friends and strangers have come by to look at what the farmer and his partner, Peterson, have wrought on this small acreage, they, too, question why one would voluntarily surrender the relative ease and affluence of a city job for this sort of martyrdom.

There are many answers, but the simplest – and this the farmer hopes all will consider as they vicariously celebrate the harvest – is that growing food is one of man's highest yet most basic callings. Now that may sound a bit highfalutin', but the farmer has come to know it is true.

Pulling that dreadful plastic, the farmer also has come to know, is one of the prices he must pay for admission to this most exclusive of circles. And in that sense, the task becomes less onerous.

Now, too much can be made of the presumed religiosity of a farmer's life and often is, usually by denizens of the city who do not know about working on all fours. But in fact, there is something sacerdotal about life in the truckpatch.

Being here and having become part of a microscopic universe in which he is a key player, the farmer has come to know that some great power guides us all and fie on those who choose to work counter to this power. This knowledge is comforting and reassuring.

The farmer cannot describe the magical aroma of healthy soil, just turned in the spring. He is at a loss to articulate the emotion of watching the first germinating seeds pushing up through the earth's crust. Without sounding terribly saccharine, he would not try to limn the joy caused by abundance in the fields or the union he feels with Nature when the crops flourish.

Nor is there a way, really, to portray the pleasure he feels when utter strangers praise his tomatoes or his broccoli. Yet, such praise has come to be more meaningful and important than any accolades he might have received for news articles he crafted in another life.

Curiously, this all came to the farmer almost by accident. Years of traveling the country, meeting inspirational figures who by example were showing how man could work in harmony with great power and be the better for it, were moving influences.

Important, too, was the realization that the best and the brightest and the most successful more often than not were people who had come to agriculture as a second career. They were not fettered by the biases and bad habits of past farming generations, which were perpetuated by large institutions or transmitted genetically. That realization gave weight to the works of Louis Bromfield, the novelist, who said that some are born farmers but shouldn't be; others are born farmers but don't realize it until they've done something else in life.

But more likely the greatest motivation for leaving the city was practicality. Peterson and the farmer tired of debating each year who would get what portion of their limited suburban garden. More space was the obvious solution.

And so, with the help and encouragement of munificent friends, it came to pass. The farmer got his space, Peterson got her space, and neither ever looked back. Not even knee-deep in dirt on a cold fall day, grappling with a chunk of plastic.

II.
THE TYRANNY
OF NATURE

LET A SMILE BE YOUR UMBRELLA

June 12, 1991

There had been no rain for many days, nor even much of a threat of rain, but the farmer had gone through this sort of perversity before and he wasn't going to let it depress him again. Day after day the mini-drought continued until it became the driest and hottest May since the farmer began keeping his records and, in fact, there was every reason to be depressed.

The plants were suffering and the ground was so dry that tillage was suspended on most beds that were yet to be seeded. Irrigation helped ease some of the plant stress but it wasn't enough to offset the record-breaking heat.

So how hot and dry was it, Johnny? Well, the farmer one day found a slug at a water spigot, holding out a tin cup. The most water in one place around here was in the cat's dish.

The farmer tried to make light and see a positive side to all of this, but his mind kept wandering. He remembered "Jean de Florette," the somber French film in which Gérard Depardieu the actor was killed in his search for water.

He remembered the old local theories for previous droughts: It was the orchardmen to the west, seeding the clouds and stealing the rain before it arrived (theory unproven). Or it was the army engineers, creating a drought to justify a new dam that would take farmers' land (impossible!). Or it was the CIA, for unexplained reasons (plausible theory).

Yet, there seemed to be something to this. Many times the farmer would see small planes fly directly into menacing dark storm clouds that soon would dissipate. Who were they? What were they up to? Feelings ran so high that other farmers occasionally fired rifles at these planes.

The farmer remembered all these things but he refused to let his mind play games this time. What was happening here, he decided, was that he was being handed an opportunity and an advantage.

The early tomatoes, blessed with a wet spring, were burgeoning in the dry with no signs of blight that often comes with moisture. The strawberries came early, but they had no rot or mold that wetness causes. The peas were free of mildew, the tree fruit showed no signs of the fungal problems that occur with dampness.

And while the lack of moisture was slowing down some crops, it also was deterring the weeds that often run rampant in late spring when the farmer's mind is focused on other tasks more pressing than keeping the beds free of intruders.

Last year and the year before were just the opposite – intensely wet Mays that caused all manner of problems. The weeds went berserk. The strawberries suffered. Slugs were everywhere, chewing up plants. Beautiful lettuces were rotted on the bottom from the wet.

So what we will do here in all this heat and dry, the farmer said to himself, is cope. Absent the rain, we will carry water in barrels to the fields and feed each thirsty plant individually. We will put blankets of straw on the open soil to hold in the moisture. We will do our transplanting only in the cool of the evening.

We will cope and we will smile about it.

A recent visitor from the Soviet Union helped set the tone. What about the food problem in Moscow? the visitor was asked. "There's no problem," he said, "because there's no food."

It's sort of like that in the truckpatch. There's no rain problem out here because there's no rain. You cope and you smile.

No More Whining About The Weather

The farmer isn't much inclined to make resolutions. But sitting here at his writing machine while a mid-March snow falls on his barren fields, he has decided to come to terms with the situation. The farmer is going to do less whining about the weather.

This surprise storm had come, and it was going to endure a while if the weather radio was correct, and nothing could be done about it.

Just live with it, the farmer instructed himself. So you don't plant the potatoes quite as early. So you don't mulch the asparagus and the raspberries. The rhubarb doesn't get weeded and the first of the lettuce transplants don't go to the field. Live with it, a voice said.

Now, this is a distinct change of approach. Ordinarily at this time of year the excitement is hard to contain. The anticipation of a new season, the sensory thrills provided by swelling buds and freshly churned soil, are things that get the farmer's motor running.

But this has been a different kind of winter and gradually it has come to the farmer — humbling though it is — that the weather is beyond his control. There's not much point worrying about it, even less point whining about it.

Down in the city, the farmer's friends commented about their mild winter and how lovely it must have been in the country. In fact, it was lousy in the country — wet, down below freezing most nights, windy, flat-out unpleasant.

The farmer's notebook reveals that there were few of those "pet" days, a local term for the unseasonably nice winter respites that allow outdoor work to get done. As a result, field preparation had been delayed and other tasks simply were ignored.

The constant wetness had made the composted manure pile almost too heavy to work. Moisture in the field made machinery use impossible. A cold wind that gusted all winter long drove the farmer to cover.

But now in his new mindset, the farmer is not too worried. He is inclined to see that there may be a good side to it all, even though he knows that his spring will be all the more hectic for it.

For one thing, the season will begin with ample subsoil moisture. This will help young seedlings get a good start. For another, the succession of lousy days has dissuaded the farmer from taking the big risks that he usually takes by planting things before their time.

For another thing, the fruit trees finally might be able to do their thing. In years past, winter pet days have lured the trees from dormancy and brought on early blooms, which then succumbed to killer frosts in late spring. Maybe this will be their year.

The enduring cold and the string of below-freezing nights will help eliminate more of the bugs that winter over and emerge to haunt the truckpatch through the summer. The new cover of snow, like the others that fell through the winter, will be a security blanket on strawberries that could be nipped by freezing weather.

As he watches the huge snowflakes tumble past the window, piling high on the distant fields and obscuring the woods, the farmer begins to grasp the wisdom of taking this unexpected blow with calm.

A friend from the next county used to watch with amusement as the farmer rushed to get ahead of each spring, putting out transplants and daring the cold to take its toll. "March is too early to get stressed out ... too early to push things," this friend would say.

The farmer would snort a bit and grumble about how there never was time enough to do all the chores without taking the risks and without challenging nature just a little.

Now, confined to the house by a storm that he cannot control, the farmer sees the value of the friend's admonitions. Everything in its time, everything in its place. Somehow, a way will be found to tend to all the demands of the farm.

Just a few days before this surprise late winter blast, the farmer had talked to his friend Musachio, who grows vegetables in warmer Maryland. "You got anything planted?" asked the farmer.

"Well, only the peas. Weather hasn't been good enough to do much else," Musachio answered with the placidity of a man who long ago came to terms with nuances of weather that drive others to distraction.

In another time, the farmer would have been depressed to learn that this friend's peas were in the ground, that a competitor might get ahead of him, that another farmer might know more than he knows.

This winter and this new snow have changed that. It is, as the other friend said, too early to get stressed out. There will be time for peas and lettuce.

The excitement of a new year is not diminished, but today, the snow falls and so be it. The farmer wishes Musachio a good season.

An Open and Shut Case

May 20, 1992

The mail one day brought a recipe for dandelion wine and so the farmer, his imagination stirred by the thought of finding a use for this pernicious invader, waited until the fields were buttered in yellow and laid his trap. The workers, city denizens possibly stirred more by the prospect of a cheap buzz, poised with their buckets after work one evening to go out and harvest the flower heads that would be converted to elixir.

Well, the dandelions had vanished. The fields that only hours before shimmered with gold were green again. Dandelion flowers were nowhere to be found. Crew and farmer were teetotally mystified. No wine would be made. Sadness prevailed.

Next morning, the fields again shone with the bright flowers and the mystery began to come clear. "Ah, so," the farmer and the crew said in unison, the flowers fold shut in the evening and unfold during the daylight.

This time there was no waiting. Flowers were collected of a morning, a batch of this legendary country potion was put to cook on the stove, and the veil of sadness lifted from the truckpatch.

The dandelion episode set the farmer to pondering some of the things he though he knew but didn't until he began focusing more keenly on what was happening around him.

Among the things he didn't know was that that selfsame dandelion made awfully good eating in a spring garden salad, having been told of this by old-timers in the country.

In time, the farmer learned that people were actually growing the dandelion green on sizable acreages in New Jersey, just to satisfy the demands of the ethnic markets. "Ah, so," the farmer said, there is money to be made with this and now the truckpatch includes long rows of dandelion greens from a domesticated variety.

For many seasons the farmer would try, and almost invariably fail, to grow certain vegetables in the spring. They worked very well in the fall, but just wouldn't thrive early in the year, usually bolting to flower before their time. The farmer tended to blame himself without really understanding the problem.

Then his friend Roland, a farmer who guards his secrets closely, came by one day with a monster of a Chinese cabbage that he displayed with pride. Roland leaned over and whispered, "I'll tell you the secret. Seed this two days after the longest day of the year and it will not bolt."

The farmer penned this vital tip into his little black book as a reminder, albeit mystified by the concept. It came clear with an ancient book's disclosure that many of these cantankerous vegetables will not bolt when days are shorter than night – born to flourish in the fall, that is.

The tip was valuable, for the farmer now no longer wastes his time trying to grow varieties that are designed to fail if planted out of synch with nature. And the burden of self-blame was put to rout.

This sort of thing seems to happen all the time. The farmer used to think that a certain pestiferous weed was a ground ivy. Cramer, a learned young apprentice, looked at the weed and decreed it to be a henbit. His assertion was checked out in a plant guide and sure enough, he was correct. "Ah, so," said the farmer.

For many seasons, the farmer regarded the wasps and yellow jackets that zoomed about the truckpatch to be pests good only for eradication and he went to lengths to get rid of them. And then he learned that these creatures feast on aphids. The farmer felt appropriately chastened and now lets them be.

Occurrences in the truckpatch also lead the farmer to revise his thinking about what he for so long had deemed to be weak and defenseless creatures of nature.

Bob the tailless cat, for example, is a relentless helper about the truckpatch, routing bothersome mice from the patches. For this he receives great praise. But Bob also loves to chase birds, which are just as important as he is to the truckpatch for their insect-eating proclivities. Bob is scolded sternly when he goes after the birds, but the fact is that he is no match for some of them.

The mockingbirds and the barn swallows dive-bomb him constantly when he goes on the prowl and more often than not drive him to cover. The fledglings survive.

On a recent day, the farmer noticed a small bird pursuing one of the five resident crows, terrible marauders that unerringly pluck vegetable seeds from the newest planted beds. The same bird would swoop in and hit the fleeing crow and finally drive it away from the seeded beds into the woods.

The next day the same thing occurred. This time it was clear that the pursuer was a mockingbird. The farmer had thought all along that this creature was put on the truckpatch only to disrupt his sleep in the pre-dawn. "Ah, so," the farmer said, "better friend than imagined."

The farmer's assumptions about the tenderness of a tiny plant are often sundered by the reality of its power. A three-inch potato plant inadvertently crushed under the heavy tractor wheel invariably fights back and lives to be a producer.

The power and force of these little plants is reaffirmed every spring when the farmer observes a just-sprouted bean or pea slowly pushing up through the soil and actually overturning small rocks as it battles to survive and produce.

And so it goes in the truckpatch, where the farmer is continually learning new things and having his tired old misguided assumptions put to rest by observation of the realities of nature's mysterious forces.

The farmer doesn't pretend to understand it very well, but he now knows that these same forces are doing their thing in two crocks in the pantry, where that concoction made from the lowly dandelion flowers will soon be ready for decanting. "Ah, so."

THE GAMBLE ON THE GREEN

March 25, 1992

If the farmer had a healthier respect for symbols, he would hang a pair of those big fuzzy dice in the pickup truck to help him stay focused on the obvious occasional folly of what he is doing. Fact of the matter is, however, the fuzzy dice might be superfluous, for it seems that hardly a day passes in the truckpatch without a risk of some kind being taken. In truth, real-life dice are rolled every day and the winning average isn't all that good.

Take the morning the farmer was awakened in the dark by the rasping sound of a snow plow out on the road. It could mean only one thing. Snow, as predicted, had come during the night and it was deep.

What hadn't been predicted, however, was the bitter cold that accompanied the snow and the farmer sensed that daylight would reveal the trouble that was sure to follow.

Just days before, about 5,000 lettuce and cabbage plants had been taken from the security of the greenhouse and placed into unheated cold frames, where they would be hardened off for transplanting in the fields. The weather had moderated nicely, and the move seemed timely.

Then came the storm.

Peterson, the partner, screwed up her courage and trudged into the snow to survey the cold frames. "They're zapped," she reported back. "Really zapped ... Not likely to make it back."

The farmer plunged into his customary funk. All that work down the tubes, he thought; all that time, wasted; all that early crop, shot.

"You might as well deal with it," Peterson counseled. "You take a risk – and you know you take a risk – every time you put those plants out there early. You're just going to lose once in awhile and that's the way it is."

Well, that's exactly the way it is, which the farmer knew without being told, and while it didn't assuage the loss of the cabbage and the lettuce, it again conjured images of dice being rolled each day.

All of which reminded the farmer of the visitor who long ago came to the truckpatch and made a solemn pronouncement after seeing what was going on. "You've got to be out of your mind to do this," quoth the visitor. "It makes no sense. It's not rational."

It was intended as humor and it drew the appropriate laugh. But it had a worrisome ring of profoundness that still comes back, especially on the morning of that March storm, to make the farmer wonder if he in fact suffers from some perverse sort of derangement.

The traditional farmers around here, the ones who grow field crops like corn and wheat, also may suffer agricultural derangement but it does not derive from early planting. They wait until the time is right and minimize their risks. They seem to share the visiting quipster's doubts about the wisdom of trying to beat the odds. It's tough enough to grow things in season, they'll say, let alone testing the weather every year for an early crop or a late crop.

Early last May, young Strait stopped by to look at a field the farmer had seeded to buckwheat for soil improvement and solemnly shook his head in wonderment. The crop was lush. "That can't be buckwheat," he opined. "Nobody plants that around here until Memorial Day. It's too early, otherwise."

Another fellow down the road scoffed when he found the farmer planting peas on a warm February day – that just isn't done either. His disapproval was unaltered by a gift basket of fresh peas in May. Another fellow impishly asks every January if the tomatoes are ready yet, knowing that they're not, but at the same time implying a sort of respect for the taking of risks.

That, of course, is what makes the truckpatch different from the conventional farms hereabouts that grow grain crops and hay for their livestock. Every week from March into September something – food for people, that is – is being planted in the truckpatch and every planting seems to carry a risk because weather conditions rarely are ideal and a hungry deer or groundhog always lurks back by the woods.

Unlike the conventional farms, the truckpatch has no federal programs to subsidize the crop, no federal insurance to cover the losses, no support schemes to prop up the prices, no futures market to hedge the bets on radicchio and raspberries.

To make it the more perverse, the truckpatch concentrates on the varieties that offer the best flavor, which are not necessarily the varieties that are most resistant to disease and weather stress. And then the farmer trucks his provender to the open-air city markets, where he runs the risks of a rainstorm or some other event wiping out his chance of selling the best lettuce of the year.

And yet there is a whole population of truckpatchers out there, going through these same feckless exercises in frustration year after year and somehow enduring and even prospering, against all odds.

So, no, the farmer is saying to himself on a frigid March morning, it isn't rational and it makes no sense. But the dice come up right just enough of the time to keep the farmer playing the game.

In a clinical world, they'd say the farmer ought to be committed. Well, the farmer is committed, but not in the clinician's sense of the word. March, after all, is not forever and a zapped cabbage crop is not the end of the world.

THE SEESAW SEASON

Feb. 26, 1992

The cold was still hanging around and the first decent snow of the winter had thrown a moist blanket over the truckpatch, yet wherever the farmer looked there were signs of nature's reawakening.

The fall-planted garlic that didn't do squat in October and November because of the drought suddenly was pushing green tips up through the cold earth. Shallots that had lain dormant were showing life signs – soil cracking open along the planted rows.

Even with below-freezing nights, the dreaded chickweed was starting to flourish and send a web of verdure across the planting beds. The Dutch iris had emerged, gotten freeze-burned and begun pushing back up through the earth.

The five huge crows that have been on duty here for two full years reappeared from nowhere. Deer so invisible all winter long had pounded hoof prints in the snow out by the strawberries. A trio of blue jays, fat and raucous strangers, appeared among the swelling buds of the lilac bushes.

Snapdragons and carnations and rosemary, experimentally left in the ground to see if they might survive the winter, had begun to show green around their bases, a hopeful sign that perhaps they would live to produce again.

Hopeful signs. These are the titillating messages a flirtatious nature sends to taunt the farmer and prepare him for the annual affair of heart with the plants and creatures of the truckpatch.

Since the farmer first began to note the ebbing of his winters, the first signs of spring always have shown at about this time, but the emotion stirred by the seasonal change always seems new and mystical.

The unseasonably balmy January and February "pet" days, as they are known hereabouts, draw the farmer from his winter cocoon and activate the juices of energy that gradually engulf the whole truckpatch. Every pet day is greeted as if it is a unique experience. So it is again this year, for these greater forces now are telling him that the season has begun in earnest, and there will be no more dawdling with coffee-drinking neighbors and no more swapping of horticultural boasts with farmer friends. Ready or not, the winter is over.

In the greenhouse the lettuce and the leeks and the thyme have pushed up their first tiny green leaves; onions and cabbage are soon to follow, and, even before the farmer is fully ready, he will be forced to take them to the fields in a matter of a few weeks.

And how or where, the farmer asks now as he asks every late winter, will the time be found to cope with the rush of tilling and planting that will not wait? Why were winter jobs left undone to now crowd up against the exigencies of spring?

In his own style, the farmer is emulating the way of nature that gives him the steadily growing little rushes of energy that say the awakening is nigh. Like nature, he will move in fits and starts toward his new spring. And when sun and moon and tides are just right, with no formal announcements, some inner metronome will begin ticking and establish a rhythm for working the fields.

That, the farmer is thinking, is probably the meaning of winter – it is a time that man, plant, and creature are allowed for sleeping late, moving at a languid pace and contemplating the majesty of what is about to come.

So it is the awakening, and even though occasional winter bluster is sure to interrupt the flow as the seasons change, the feeling is good, and the optimism rises as high as it does every spring.

HOPE SPRINGS ETERNAL

April 18, 1990

In the country, whether there's anything else to talk about or not, no conversation can occur without some discussion of the weather. And so it was that Donny the pump repairman had a thought.

It was one of those 90-degree days in early March – what the weatherman called unseasonably warm – and Donny reported with some astonishment that his television had shown Washington's cherry trees in full, unseasonable blossom.

"You'd think they'd learn down there," he said. "Cover the trees or something."

The farmer explained that for all its other faults, Washington could not be brought to book for the early blooming cherry trees. But on reflection, the farmer thought, perhaps by now he himself would have learned.

The lesson, of course, is that the weather cannot be counted on for anything but surprise and heartbreak. Thus the farmer must not allow his excitement and his impatience to make him plant things before their time. Yet he keeps ignoring the hard-learned lessons.

Take the peas, for example. Those seductively warm days of early March told the farmer to get out there and plant the peas. Into the ground they went and then the weather turned nasty cold and the seeds just lay there, knowing the farmer had done them wrong.

The tender, tiny seedlings popped through the ground 10 days ago and then got blitzed by more cold. The only thing that saved them was a blanket of snow, which acted like a layer of insulation.

Maybe they'll make a crop, maybe they won't.

Or take the potatoes. Following tradition, the farmer likes to put them in the ground by St. Patrick's Day. Plant them early, dig them early, and destroy the competition, the thought goes.

Well, when the farmer talked to Ronniger, his potato supplier in Idaho, he was warned. "You surely don't want them in March," Ronniger said. "If they freeze, they're lost. We won't even ship them until the danger of a freeze is past."

The farmer agreed, but he couldn't resist. It was March, by gum, and he had to have potatoes in the ground. He went to the Dott store and

42

bought 150 pounds of spuds. The storekeeper asked if it might not be too early to be planting. "No," the farmer said, "you've got to take your risks."

Then Fowler, the county agent, called. "They're talking about 20 degrees tomorrow night ... I sure wouldn't plant those potatoes now," he said. "There's just too much risk of loss if the ground freezes."

Fowler, thankfully, intimidated the farmer into waiting. It hit 20 degrees and it stayed there. The ground was like a rock. And the potatoes waited until the end of the month before they were planted.

The fruit trees, just like Washington's cherry trees, were another matter. The warm days of February and the record heat of early March lured the peaches and apricots into bloom. And when winter came back and settled in for awhile, those lovely pink blossoms blackened. Most of the crop was sure to be lost.

If the farmer were truly serious about his tree fruit, he would do as the prudent orchardists do. He would install an overhead spray system and, with the threat of a freeze, turn on his sprinklers and create an insulating coat of ice around his trees to save the fruit. Yet he hasn't done that.

It's not that the farmer hasn't made some effort to deal with winters that seem to be increasingly temperamental and inconsistent. To avoid early blooming, the farmer put his trees on a north-facing slope so they'd remain cool and not be tempted by premature warmth. And he stacks mulch around the trunks to keep the root area frozen and make the trees less vulnerable to early warm days.

Good idea, nice try, but it basically doesn't work. You'd think the farmer would learn.

Same thing with the berries. Beguiled by the memory of a long-ago mid-March when the weather was delightful, the farmer continues to have his new berry plants shipped in mid-March when, it turns out, they almost never can be put into the ground.

Strawberries, raspberries, and blackberries arrived on the exact day in March for which they had been ordered, and, predictably, the weather was too miserably wet and cold and windy to be outside planting.

Finally, the tender plants could wait no longer. So raspberries were chunked into the ground on a bitter cold day in April and the strawberries and blackberries were planted two days later when it was only marginally nicer.

They might work, they might not. The farmer is learning. Next year, he has resolved, the berry plants will not be ordered before their time. Unless the farmer gets impatient, that is.

Those unseasonably nice spells in January and February twist the farmer's mind in other ways and make him do other things he shouldn't do. One of them is rushing to the greenhouse and starting some of his early plants before it's really time. They do just fine for awhile in the poorly heated greenhouse and then they get stunted or zapped by the roller-coaster temperatures that are beyond the farmer's control.

That means the farmer must keep one ear on his weather radio and one eye on the greenhouse. A cold warning sends him scrambling to the greenhouse each evening and forces him to move dozens of flats of plants to protected areas. Each morning, the procedure is reversed.

More work, more frustration. But it's become such a part of the routine that the farmer's daily work list almost always has a line that says "schlep trays," meaning tend to the greenhouse at night. There again, you'd think the farmer would learn.

Then, miraculously, the plants somehow reach a stage when they must be taken out to the fields and transplanted – good weather or not. Broccoli, cabbage, and rapini all reached that stage last month and with the first break in the cold, hundreds of them were put out under lightweight plastic protective covers.

And there they remain, not growing but not dying either, suffering no doubt from the continuing chill and sunless days, undecided and giving no hint of the June harvest the farmer had in mind when he put them in the soil.

Radishes, scallions, and lettuce transplants are in the same boat. They all were planted on those misleadingly springlike days of March and they all sit out there in their beds doing essentially nothing. If the cold and the damp don't get them, maybe the crows will.

The annual quandary over spinach – will it come up? will it survive – was settled early. A band of huge crows descended on the spinach beds and lingered for several days. A closer look revealed jillions of tiny holes along the seeded rows, each hole the size of the crow's beak that had extracted the spinach seed from its place.

So you'd think the farmer would learn. He would get a heater for his greenhouse, he would restrain his urge to plant things too soon, he would become immune to the false weather signals, he would get some sort of gun to dispatch the marauding birds and he would buy himself some peace of mind.

But he doesn't, and there's no good explanation for it. Deep in his heart, he probably enjoys the misery. Or it gives him something to talk about when Donny the pump repairman comes along.

How To Make a Molehill
Out of a Mountain

July 11, 1990

Less than half a mile to the west of the farm is a high ridge, distant enough to be imposing yet near enough to abet the sense of coziness here, and it is on this ridge that the farmer's eye is riveted each day.

The farm's fields dip and roll, sometimes precipitously, but there is no point from which the ridge cannot be seen. The ridge dominates physically, but it now also has become a psychological fixture.

On hot summer afternoons, when the sky darkens many miles beyond the ridge and when the faint boom of thunder can be heard, the farmer studies this huge natural barrier more intensely and even talks to it.

The issue always is whether the storms will be strong enough to push over the ridge and bring merciful rain to the parched beds in the truckpatch. Or, conversely, whether the ridge will be generous enough to allow the rain to move on through.

Precious little rain made it past the barricade in June and the farmer has taken to blaming the ridge.

On more days than not, the heavy black clouds would gather behind the ridge and then simply disperse. The farmer could look far to the south and see the sheets of rain driving down. He could look far to the north and see a similar picture.

So the farmer sometimes imagines that his ridge is an enormous dam, placed strategically to hold back the water that might come with the thunderstorms that seem to fall everywhere but here.

Other times, the farmer imbues the ridge with human qualities, imagining that it somehow has become a judge and disciplinarian, holding back the rain clouds as punishment for some miscreancy.

Some nights, deep in his sleep, the farmer dreams about rain and floods. He sees withering plants resuscitating from torrents of water and he sees browning fields turn green again.

These are games that the mind plays when the need for moisture is desperate. It shouldn't be this way, but the farmer never quite comes to terms with the obvious fact that he is powerless to change the weather. He never quite understands that he has no recourse but to accept his lot.

Not that the mind is acting alone.

The farmer relies on the federal government's weather radio to guide his daily routine. He listens closely for the "thunderstorm potential statement," as the weathercrats like to call it, and his hopes rise and fall accordingly.

Through much of June, the radio was predicting the afternoon showers that were needed so badly. But they did not come. The farmer blamed the weather service first and then he blamed the ridge. He sometimes sensed a conspiracy. He even came to believe the local consensus that an unidentified government agency manipulates the weather in order to drive farmers off the land.

The truth is — and the farmer's journals reflect this — that June and July more often than not are dry months here and nothing can be done about it except to turn on irrigation lines that reach only a portion of the farm. To the untrained eye the appearance of green around the truckpatch might suggest that all is well. In fact, however, the prolonged dryness affects the tender vegetable plants at a critical time for maturing.

Going through the journals as July came, always with one eye cocked toward the ridge, it dawned on the farmer that almost without realizing it he had entered into the annual dry season and that his year was half over.

The farmer equates this time to baseball's All-Star break, the midway point at which the players pause, catch their collective breath, and gird themselves for the final hard drive for the pennant.

It is similar on the farm.

Imperceptibly, the spring crops have come and gone. The lettuce had flourished from decent early rain but was petering out with the heat — as always. Broccoli, which did well, had been cut and the fall crop of seedlings was growing apace in the greenhouse. The snap peas were finished and the green beans were in full flower, although desperate for water.

Almost without noticing, the farmer already had undergone the excitement of searching out and eating the first of his June-ripened tomatoes, tastier from the heat of the field, and he had thrilled at the sight of peppers and eggplants forming on the bushes.

Almost without notice, squash and melons – at least those that had survived the cutworms and the seed-robbing crows – were pushing out into the spaces between rows. The pumpkins were doing passably well; the potatoes were forming nicely in their hilled rows.

All of these were signs of the farming year having reached the halfway stage and, truth be told, it hadn't been all that bad so far. It had gone so quickly that the farmer was caught unaware.

The thought occurred that perhaps he had been paying too much attention to the ridge, affixing too much blame to its perfidy, and perhaps he was paying not enough attention to the business at hand.

All of this he knew, yet this morning, with the first light of dawn, the first thing the farmer did was look up at the ridge to see if it was still there and to search for a rain cloud.

The mind games began again and it was easy to forget that this was July and nothing would change that. The ridge was still there, not a cloud could be seen, and not much else was new.

EVERY SILVER LINING HAS A CLOUD

Aug. 19, 1992

There was the Habanero, the hottest thing known to man; the tiny Thai pepper, the chunky jalapeño, the Hot Portugal, the cayenne, the Super Cayenne, the Español, the serrano, and an awesomely destructive thing of unknown lineage named Jared's Hot Lips, after the worker who provided the seeds long ago.

Once the season got going in earnest, the farmer would slip away to a distant field every few days to check on the progress of what he imagined might be called a crowning achievement. In that patch has been planted more than 600 hot peppers – enough culinary incandescence to light up a small city – in a wondrous array of varieties intended to satisfy every taste.

The plants flourished as never before because it had been that kind of year out here – not too much rain, but just enough; not too much heat, but just enough; not too much glaring sun, but just enough.

The farmer found it difficult to contain himself. If these peppers were searingly hot in other years when growing conditions were not as good, imagine what this crop would be. Something like Hell in a salsa dish. "Killer peppers," Cramer the chef would call them.

As the fledgling peppers formed, the farmer would randomly, and always tentatively, sample them. Some were hot, some were less than hot, some were on the verge of getting there. They were slow to warm up, the farmer reasoned, because the growing year had been slow.

So, having heard these periodic reports, the farmer's friend Cramer, who intended to make spectacular use of this crop, found it just as hard to contain himself. He would gingerly finger the last remaining Thais of 1991, dried to carry him through the winter, and expectantly rhapsodize about the new peppers.

"Killer peppers," Cramer would mutter. "I want killer peppers. When will they be ready? Where are they?"

Finally the first picking was hauled to Cramer, who was already rearranging his menu to accommodate the zest. Cramer was exultant. The farmer, pleased, sat back and awaited reports of major fires in the city.

A week later, Cramer was not a happy man. "What is it with these (expletive) peppers?" he groused. "After all this fanfare you bring me peppers that aren't hot ... zilch heat, nil, nada. What did you do wrong?"

Well, what the farmer did wrong was assume that since he was in the midst of a beneficent growing season, his hot peppers would automatically be the best to ever come from the truckpatch. The farmer took the chef's displeasure personally. Peterson, the partner, took the criticism in stride.

"This is why hot peppers are grown in New Mexico," she said. "The heat and the dryness make the pepper hot ... This just may not be the year for hot peppers in Pennsylvania."

Which the farmer was beginning to understand. The point of the pepper story is that the farmer again is reminded that things rarely are what they seem in the truckpatch. The plants and the seeds from which they spring actually know that better than the man who nurtures them.

Here in the middle of what is generally regarded as a "good" year – at least good as the human defines it – the verity of this reminder seems to leap from every corner of the farm. Some plants would just as soon have a little stress, a little angst, to do their thing properly.

How else to explain the peppers? Last year, with the worst drought in a century, hot peppers with no irrigation were seriously hot. Sweet peppers without irrigation were delectably sweet and full of flavor, borne heavily on plants showing no sign of disease.

But this year, in this apparently best of years, the peppers are in a full pout. As the days recently have warmed, the hot peppers are only now beginning to strut their stuff. The sweet peppers remain indifferent. The plants are gaunt and not at all impressed with the rain.

The tomatoes that tasted so great in a drought year – the heat intensifying the flavor – have been slow and balky in this year of rain and cool spring. The plants are fine, lush and healthy, and their fruit abundant, but the taste just isn't there.

The Early Girl tomato, one of the best tasting, is but a shadow of its usual self. The Pineapple tomato, an antique variety that was unspeakably sweet last year, can't seem to figure out what it wants to taste like this season.

And so it goes as this growing season pushes toward fall. The plants – tomatoes, eggplant, okra, squash, and others – have set abundant fruit all of a sudden and seem driven to complete their mission. It is as though they know something not yet revealed to the farmer about the coming autumn. As though they want to get done with their work and just be out of there.

The farmer, too, noted earlier in the month a subtle change in the light and the air, as if fall's advance man was making the rounds, but he still isn't sure what to make of it.

These plants are trying to tell him something, if only he could learn to listen and figure out their language. If he achieved that, it could be a very, very good year. A crowning achievement, even.

III.
THOSE WHO
SHARE THE LAND

A PLACE TO LIVE AND LET LIVE

Oct. 2, 1991

O n one of his late evening patrols, when the shadows were long and the September sun still provided a balm, the farmer was stunned to see a squadron of tiny helicopters over the cauliflower.

Hither and yon they cruised, just above the tops of the plants, on what appeared to be a purposeful mission. These were praying mantises, in numbers not before seen here, and they were vacuuming pestiferous bugs from this corner of the truckpatch.

Now, mantids have been here for years, but this was unusual. The sight of so many caused such wonder in the farmer that he could only stop and watch in awe this display of nature once again negotiating an ecological detente without the help of man.

The driving purpose of the truckpatch, of course, is to produce food for people in the city. And to do that successfully, the farmer realizes increasingly with each passing season, these contending forces of nature must be left alone to work out their differences.

The apparent sudden burgeoning of the praying mantises got the farmer to reflecting on a reader-friend's recent curiosity about the "hidden nests" around the farm and how they affect the growing of food.

The farmer encounters these "nests" almost every day and almost every day he is impressed or moved by these signs of clandestine activity in the wild. He occasionally wishes desperately for the magic powers of night vision so he could drink it all in.

For over time, notwithstanding a certain smugness about the imagined firmness of his command here, the farmer has come to understand that he is not alone in the truckpatch. In fact, he may be no more than a bit player in this dramatic kingdom of nature.

That clump of crown vetch, for example, was high and bushy when he last looked at it. This morning it is indented with the shapes of recumbent deer who obviously spent the night and then moved on unseen to other warrens.

The tall grass down by the cabbage beds was undisturbed yesterday and yet this morning it is cut by a thin trail, a path blazed by a groundhog in search of food. The trail begins in a culvert, at the lip of

53

a burrow previously unseen by the farmer, and ends at a nibbled cabbage plant.

It is easy to denounce and detest these predators – the deer, the groundhogs, the rabbits – and to take dire steps to thwart them. But the truth – and the farmer knows it – is that they were here before he was and only with difficulty will they be thwarted.

The farmer feels good about the new electric Maginot line he has erected to protect his tender crops from the deer and there are signs that it works. Still, the farmer this morning finds hoof prints inside the barrier and he knows he is defenseless.

His "humane" groundhog trap is a wonder, having snared more than a dozen of these marauders since July. But today he finds that newly blazed trail to the cabbage and tomorrow his tractor will shudder as it hits yet another hidden burrow in another grassy patch.

And so it is.

The farmer reaps no joy from these discoveries. Nonetheless, the balance sheet is positive, for the surprises of each day bring a new sense of excitement and contentment from a sort of chaotic harmony in the truckpatch.

Here, on a barn rafter, is another new mud-daubed swallow nest, a sign of new visitors in search of insects. There, atop each of the stakes holding defunct tomato plants, sits a bluebird. Dozens of bluebirds, resting between bug sorties.

From down in the woods last night came the shrieks of a screech owl, peals of mayhem it seemed, to let the farmer know that in fact all was well. This morning it was nature's alarm clock, a yapping mockingbird, announcing dawn in the lilac bush outside the bedroom.

In the raspberries there are billions of bumblebees, oblivious to the humans regularly gathering fruit, doing their work of pollination. A visiting professor voiced surprise, wondering why the bumblebees by far outnumbered the honeybees. Don't know, the farmer said, they're just there.

These same bees are blanketing the flower fields, bothering neither the harvesters nor the droves of butterflies that course back and forth among the blooms. As the truck hauls cut flowers back toward the cooler, the bees and butterflies swarm after it seeking a last sip of nectar.

In other parts of the truckpatch, jabbering killdeer rush about and throw conniption fits at the sight of intruders who may unwittingly trample their hidden ground nests. Elsewhere, an unfamiliar sharp-tailed grouse is spotted for the first time, sending the work crew to the bird guide for more information.

In the beds of parsley and fennel and dill a wondrously adorned black-and-lime creature nibbles on the greenery – a dreaded parsley worm working its evil will. The workers express alarm about the threat but Peterson, the farmer's knowing crony, is not fazed.

All the expert books warn that this pest must be dealt with summarily, but Peterson gently removes these worms and transfers them to nearby weeds. She feels too strongly about the black swallowtail butterflies that these worms will produce to exterminate them.

So, the dreaded parsley worm is alive and well, but so are the parsley and the fennel and the dill. There is room enough and food enough for us all, the farmer reasons, and there is cause for joy.

Praying mantises, bluebirds, deer, parsley worms, farmers; we're here as a family of sort, leaning on each other for support, because that's what families do and because, most of all, we need each other.

SPRINGTIME WAR GAMES

May 3, 1989

It wasn't the best of springs, being either too cold or too wet to get on with the early planting tasks, and the farmer was morose. His work was delayed and his frustrations built.

So it became a duel with nature. When the days were tolerable, the farmer worked outdoors in fits and starts. When the weather was inclement, he puttered with seedlings in the greenhouse and he stewed.

But all the while a sense of foreboding grew. Tilling and planting were off schedule, the rhythms of labor were disrupted. The farmer knew that inevitably all hell soon would break loose and he would be strained to keep pace. More stewing, more frustration.

The first sign of trouble came in early April with the appearance of a cabbage butterfly, a white wraith that leisurely flitted about a sun-speckled meadow. And as the days warmed, more of these seemingly harmless creatures would show up on the farm looking for an early season free lunch.

As their name implies, cabbage was their quarry. But since there were no cabbages planted here, any other member of the crucifer family would do just fine – a broccoli, perhaps, or a cauliflower. Or a kale plant, a turnip, a radish, of which many were in the ground.

The butterflies' eggs, laid on these plants, would hatch into soft velvety worms the color of a double-lime daiquiri and the destruction would begin. The farmer's 4,000 broccoli plants would be munched to shreds unless steps were taken to deal with the threat.

Combat between farmer and insect has gone on since man dropped his first seed into the ground, and man, of course, has not always prevailed. But modern science has given the farmer a vast arsenal of weaponry to deal with the predators that can ravish a field or an orchard overnight if left unchecked.

The most common of these weapons are pesticides and insecticides, which for good reason have acquired a dark image in recent years. They are products designed to kill – hence the "cide," as in homicide, genocide, or suicide. The dark image derives from increasing evidence that many of these compounds are as threatening to farmers and consumers as they are to the bugs they're intended to eliminate.

Thus there is great debate over the safety of the American food supply. How much residue of a pesticide is tolerable? At what point does the use of these lethal products outweigh the public health risk? What unknown but surely ominous forces are loosed when these pesticides combine? What makes farmer and chemical companies so certain about the safety and reliability of these poisons?

Fortunately, in this truckpatch, at least, the questions are not germane, for science and horticultural lore also have given the farmer other, more benign ways to deal with his bugs. In the vernacular, he is an "organic" farmer, which means among other things that he simply does not use the popular synthetic killers of insects.

The challenge for the farmer who adopts this approach is to work with – not against – nature. He knows that healthy soil makes healthy plants, which are anathema to many bugs, so he strives to upgrade the condition of his fields. He knows that white-hat bugs, if you will, can be put to work against the black hats. He knows that certain plants contain mystical powers that will repel the predators.

In the case of the cabbage butterfly, for example, the farmer has several options. He covers some of his crops with a gauze-like fabric to stop the darting flyer from reaching the leaves of the crucifer to lay its eggs. When the plant has matured, the cover is removed and a spotless vegetable is sent to market.

Or, if the crop is left uncovered and the cabbage worm appears, the farmer applies something called Bt., short for Bacillus thuringiensis. As the worm ingests this bacteria that is sprayed on broccoli leaves, its digestive system is paralyzed and the threatened plant is safe. The Bt bacterium is lethal only to caterpillars.

The organic or "biological" farmer uses other natural poisons to deal with other pests – rotenone, derived from the root of a tropical plant; pyrethrum from a type of daisy, and sabadilla dust, produced from the seeds of a South American lily. He repels some pests with an insecticidal soap made from fatty acids.

For the consumer, these compounds have a decided advantage. They break down quickly in the environment and leave no minute residues on the food. But they are still pesticides and the farmer must use them with care. Unless he waits until dusk to spray, for instance, the farmer could wipe out his hard-working honeybees.

To eliminate the Mexican bean beetle from his green beans, the farmer imports the microscopic pediobius wasp, a white-hat predator. He uses milky spore and pheromone traps to dispatch his Japanese beetles. He catches fruit flies with a sticky substance applied to plastic "apples" that are hung in the trees.

There are other devices passed down in the lore of the garden. A dusting of crushed limestone helps deal with the worst of the worst, the Colorado potato beetle. A dusting of wood ashes, gathered from helpful neighbors, sends cucumber beetles and squash bugs scurrying. A home-made spray of garlic and crushed hot peppers thwarts some pests.

The farmer also resorts to "trap" crops – small plantings of certain crops that divert a bug's attention from plants of economic importance. Quite serendipitously, for example, he learned he could keep pestiferous flea beetles from the broccoli by planting a small bed of Chinese cabbage that perfectly satisfied the bug's taste.

And for reasons entirely unknown to him, the farmer discovered he could repel bugs from his cucumbers, melons, and squash by dropping radish seeds between every plant. A little bit more work than dousing a field with a potent pesticide, but cheaper and safer.

Every year, new knowledge and new techniques come the farmer's way, through experience, through the wisdom of century-old farming manuals, through trading views with other growers who seek the magic of putting nature to work to deal with the problem of pests.

Just as the man-made toxins are imperfect killers, not even the best of the natural techniques is perfect. Once in awhile, a bug makes it through the tightest defensive line the farmer can erect and, oddly enough, it is sometimes reassuring.

Last year, a young woman toting a baby in a back sling came up to the farmer's stand in the Takoma Park market and filled her shopping bag with fresh-cut broccoli. She extolled the broccoli, but stunned the farmer.

"I just want you to know I found a w-o-r-m on the broccoli I got from you last week," she said.

"Oh, my, I'm embarrassed," the farmer said. "I'm really sorry."

"Don't worry about it," the woman said. "At least I know you're not using those awful pesticides." Nobody, least of all the farmer, wants a worm on his broccoli or a scar on his apple. But considering the alternative, it's not a bad trade-off.

A SENSE OF BALANCE
WITH THE WILDLIFE

May 15, 1991

There seemed to be moisture enough. And even though coolish nights and sunless days were slowing the progress of the early planted things, the farmer didn't complain too loudly. The farmer has concluded that the plants know more than he does and when conditions were right, they would burgeon.

If the peas and the lettuce and the cabbage, the beets and the other things could be kept free of weeds and other intrusions, all would be well.

Then a hammer blow struck.

Peterson, the cohort, marched into the kitchen one evening after a trek through the fields. "You don't want to hear this," she said. "The deer have hit the broccoli."

This has become ritual. At several points every year the deer will sweep through the truckpatch and decimate a crop or two. Damage is severe and both time and money are lost. These creatures may be distracted temporarily, but eventually they find the lode.

So the farmer, increasingly inured to these blows, did not even swear on this evening of bad news. "How bad is it?" he asked.

"Bad," said Peterson.

Deer tracks were all over the beds, as if a great cotillion had been held out there. The tops of perhaps 700 healthy little broccoli seedlings had been nipped off. But it was worse – they also had ravaged the newly transplanted Russian kale that lay adjacent.

To try to save the rest of the earlier-planted broccoli, Peterson and some helpers went out in the dusk and tied rags to stakes and then doused them with after-shave lotion, a short-term step that often thwarts the intruders.

So far it has worked, but deep inside the farmer knows he's only buying time. The deer or the groundhogs or the rabbits will be back to hit the broccoli, or more likely, something else just when it is least expected.

The frustration runs so deep around here – some farmers have stopped growing corn completely because of the damage – that the only remedy is to laugh about it.

A neighbor likes to joke that he spies deer wearing little bibs, poised along the roads at sundown as they decide which crops to scarf up. Peterson imagines a voice in the dark out in the lettuce patch calling, "Bambi, your table for four is ready..."

The farmer may smile, but he doesn't find it funny, for he is not running a philanthropy. Even the cats who lie around in the sun have their vital chores. Thoreau reveled in the idea of sacrificing part of his bean crop to marauding woodchucks – they had to eat, too, after all – but then Walden was not a truckpatch.

What the farmer gleans from this is that he has to come to terms with his surroundings and adjust himself to the habits of intruders that nature constantly sends his way.

A good example is the killdeer, a spectacular bird that comes here in increasing numbers each spring. This is all to the good – and encouraged, even – because the killdeer is a lusty consumer of bugs and weed seeds that plague the vegetable farm.

In return for this service, the killdeer asks for some space and some respect. Since the killdeer is a ground-nesting bird, the female invariably drops her eggs on beds that the farmer has scheduled for tilling and planting.

And as long as these nests hold eggs, usually for the better part of a month, the farmer's work is stymied. This year the killdeer took over areas ticketed for eggplants and potatoes and proclaimed squatters' rights.

Once the birds are spotted and once they begin the maneuvers that are designed to draw man away from the well-hidden nests, the farmer knows that he must move with caution in these fields. Once found, a nest is watched carefully until the babies take flight and the "all clear" sounds.

Well, the downside is that only now in May are these eggplant beds being readied belatedly for planting and only now are these potato areas being prepared. The delays are unfortunate but not fatal, the farmer reasons, and there still will be ample time to make a crop.

But there is this. The more the killdeer finds the truckpatch an inviting place, the less the farmer need worry about invasive damaging insects. The killdeer will eat whatever's at hand – a grub, a beetle, a cabbage worm, a grasshopper – and that is a contribution.

None of which solves the deer problem, but it helps the farmer cope and gives rise to the thought that one of these days a balance will be achieved on this small tract between man and nature, each aiding and abetting the other.

The farmer wishes his neighbors no ill, but he keeps hoping against hope that the fellow across the fence line will go back to planting corn or other succulents that will attract the deer. Let someone else set Bambi's table.

CREATURE COMFORTS,
GREAT AND SMALL

——————

May 16, 1990

The farmer was in a field setting tomato stakes, in deep grief over the loss of a friend, when nature of a sudden reached out with a series of gentle and reassuring touches. Not far from the tomato plot, a pair of killdeers flitted about a nest on the ground where they guarded three tiny brown eggs. As the tomato stakes were set, brilliantly colored bands of goldfinches and bluebirds came to perch atop each pole.

Overhead, barn swallows and tree swallows – and there seemed to be more of them now – darted in search of insects. In the distance, from the woods edging the field, a lonely pheasant periodically called with his sound of a steam calliope. A woodpecker pounded away at a dead tree.

Two cats named Bob and Tigger, appreciative strays who came to be part of this tableau, dawdled at the farmer's feet and followed him about, seeming to understand the miracles that were happening here. And later that day after the first batch of peppers was planted, rain came exactly as the farmer had wished.

These are some of the compensations for life and times in the truckpatch that are often stressful, always tiring, and sometimes depressingly frustrating. The farmer derives as much joy from these intangibles as he does from his customer's plaudits.

In a sense, the bounty of the farm is as much for beast as it is for man because that is part of the design of things. The birds that have shown up here in astonishing numbers are part of the work force, constantly patrolling for insects harmful to the vegetables. And on blue days, their color and their song uplift the farmer.

They even bring humor to the truckpatch. As the farmer drives his tractor in the fields, the swallows playfully circle and dip ever closer, trying to nip his head. The bluebirds follow the tractor and even seem to smile when the farmer talks to them. A mother killdeer once blockaded the tractor's path so her nest would not be destroyed.

Some years ago a neighbor came to the farmer with an anguished face. The poison ivy and brush along an adjoining fence line had become so intense that he wanted to kill it all with a chemical herbicide and then bulldoze the whole mess into a hollow.

No, the farmer said, the brush line is desirable because birds are wanted here. The farmer proposed instead that he would clear the fence by hand – on both sides – and keep it clear. The neighbor was hurt but accepting, for he had no choice. They even became friends.

Another neighbor, a lawyer, wanted to clear out the farmer's woods to get the valuable timber before the gypsy moths did. The farmer said no. He didn't need the money and, besides, he valued the whucka-whucka-whucka sound of a woodpecker hammering on a dead tree.

So the birds have come and the design is working.

The goldfinches, scores of them it seems, feast on the dandelion seeds that threaten to overtake the farm. Mockingbirds, perhaps attracted by the berries of the poison ivy, have come in droves. Barn swallows, left to foul the outbuildings with their messy nests, repay the farmer by sweeping his fields for bugs.

There is a certain thrill in this beyond the practical. One summer day, it is the sighting of three snow geese, nomads who had never before stopped, busily grazing in an oat field. Another day it is a wandering heron in a tall pear tree, or the honking of ducks lumbering north in erratic formation signaling winter's end.

And if the sounds and sights lift the farmer when things get low, he is sustained as well by smells and tastes of the truckpatch.

In the spring, his nostrils leap at the first sweet whiff of the new-mown clover. His head swims with the aroma of healthy soil being plowed from its winter rest. He savors the smell of lavender and basil and bee's balm cut accidentally when his mowing machine gets too close.

In the summer, the farmer has privileges known not even to kings and presidents. The first mature broccoli he eats raw in the field. The first few ripe melons and strawberries go not to market but to the farmer's tongue. Sun-ripened tomatoes, always more tasty in the warmth of the field, are there for the taking. The sweet corn is best eaten raw, just ripped from the stalk.

In the fall, the farmer is overwhelmed by the profusion of smells – if not over-ripened melons and extra-sweet fruit, then the special perfume in the air after first frost. And always, of an evening, the faint and inoffensive smell of manure when a distant neighbor opens the barns and releases his cows after milking.

Some of the farmer's friends, not privy to these secrets of the truckpatch, worry that his work may be too burdensome, too dreary, or too depressing to hold his interest. Not to worry, the farmer answers, for the pay is greater than anyone knows.

THE IMPORTANCE OF BEING BOB

June 17, 1992

The rain has been generous and the crops are growing abundantly if slowly, but the truckpatch has not been a happy place these past few days, for a good friend is gone.

The last thing Peterson said to him was, "Come here for your morning maul." The last thing the farmer said was, "Bobby, get your butt off of that road."

And moments later Bob, an extraordinary personality who had been sent here in the disguise of a cat, was blind-sided by a motorist who didn't even bother to slow down.

For a long time, the farmer had been unable to grasp the meaning of pets to a farm. He didn't like the idea of pets, at least in the city where they are cruelly cooped up, and he would chuckle smugly at others who developed manic attachments to their animals.

When he came here the farmer was puzzled to find cow pies around the porch and then amused to learn that his predecessor, a dairyman, had kept a cow that was more than 20 years old around the place because it made more sense than sending it off to become hamburger.

Then young Bob came along and he helped the farmer understand why every farm needs animals. Bob took over certain jobs, which he did with great skill and reliability, and he helped the place run more smoothly.

Bob was introduced to Truckpatch readers in May of 1990, when the farmer wrote about nature's creatures easing the hurt of the death of a friend. Bob and his mentor, Tigger, followed the farmer around the fields that day, seeming to know that this was a tough time and that their presence was needed.

Bob appeared again in the column last month, described as a "relentless helper" whose remarkable mousing skills far outweighed his penchant for chasing the birds whose help is also needed here.

A young worker asked Peterson what conclusions could be drawn from Bob's demise. She said, "Bob's life was perfect; he had all the comforts. The message is to go out and live life to the fullest, do what you want to do, for there's always a log truck around the corner."

Ironical, that, because Bob showed up here four years ago bloodied and bruised after he had been tossed from a passing car. He was gray with little white boots and he had no tail, which made him look more like a bear cub than a cat.

If the handicap of no tail made a difference, Bob never let on. He carried himself with the air of a self-confident individualist who was completely comfortable with his station as confidante and helper in the truckpatch.

His day usually began by awakening the farmer before dawn. He would perhaps grab a bite from his dish and then head outside, where he ran his traps. He would go from building to building, then over to the orchard, across to the workers' cabins, and back to the house.

Bob's scratch on the door was predictable – almost always at 7 a.m. sharp. He would come in and go directly to the bedroom, where he would tug at Peterson's hair and nibble her watch. He was very good at wake-up calls.

During the day, between his naps in the barn or the greenhouse (he had his own entrance, a hole torn in the plastic cover), Bob would go to the fields to help with the transplanting or to accompany the crew as it weeded onions or whatever.

Evenings, when the weather was fit, he'd scratch at the door to tell the farmer and Peterson it was time for the ritual Long Shadows Walk. Bob seemed to know how important this stroll was and he knew his job was to push the farmers, no matter how weary, to accompany him on the inspection tour.

So Bob came to be a key part of the truckpatch team and while his English wasn't all that great, he acted as though he understood the farmer's little flights of fantasy that amused workers and neighbors but left them wondering about his sanity.

The farmer's story was that Bob was valedictorian of his class at the high school. He didn't go to college, but he could have. Bob's ambition was to matriculate at Penn to study veterinary medicine, with a small rodent specialty.

But remedial math problems kept him out of school and he stayed back to help on the farm. He talked about doing brush-up work at a junior college, but the farmer couldn't spare the pickup truck to get him there. Besides, Bob couldn't see over the steering wheel.

Instead, upstairs in the home office, Bob worked on a novel, using the tiny laptop computer the farmer got him for Christmas. It had a special keyboard and it typed characters in Cat that only Bob could read.

Once, probably spurred by the ennui of winter, the farmer nominated Bob in a local TV station's Student Athlete of the Month contest, signing Peterson's name instead of his own. Peterson didn't think it was funny, except she knew in her heart that Bob deserved the award.

Well, the television people never called for the interview and that is their loss. Bob never reacted, one way or the other.

So the Bobber-Dobber, as the farmers called him, has gone to rest beside Tigger, his old buddy, where sprigs of catnip and a bouquet of flowers mysteriously appeared.

Bob's departure left some sweet memories around the truckpatch. He also left an unfinished novel, which would have been quite a read.

THANKS FOR CREATURES
GREAT AND SMALL

—————⊷•⊶—————

Nov. 18, 1990

The first killing frost came later than usual this year and the farmer, longer in the tooth but not all that much wiser, whispered a quiet thank-you. It had been a good year from almost every standpoint, abetted by rain at most of the right times and not too much heat, but the farmer now was ready for it to end.

The tomatoes had lost their taste, the peppers were sizing down, the eggplant stopped stirring excitement, and the cut flowers were fading fast. These plants knew, better than the farmer, that the time neared to pack it in for the winter.

In years past this was always a painful time for the farmer. He wanted just one more week of flowers, just 10 more bushels of tomatoes and one final flush of sweet basil. And then one more. He would tell friends that he was praying for frost to relieve his misery, but most of the time he said it in jest.

But this year was different. The farmer's serious outdoor growing season is defined by the last frost of spring and the first frost of fall. The last came on April 19, the first on Oct. 20, which made for an extraordinarily long season. The farmer was ready to stop.

The work seemed to go on forever and the plants kept producing. The abundance was so great and the time to gather it so short that for the first time, some things simply stayed in the fields. The farmer mourned, but events were beyond his control and he learned to become more accepting of the reality. He was glad to see it end.

Now, driven from his fields by the cold, the farmer begins to understand how right the early settlers were in putting aside a day to celebrate the bounty of the fall and to give thanks for it all.

The practice endures, of course, albeit in a new context. As we have become an urban nation, we continue to give our perfunctory thanks but many of us have lost sight of the reason for celebration.

Here in the country, the farmer each year seems to acquire a larger sense of gratitude for the gifts of the soil and the seasons. He congratulates himself for the successes, often blames external forces or his own benightedness for the failures.

In truth, though, he has come to know that powers greater than he regulate his success. His satisfaction comes from knowing that these powers accept him as a partner and encourage him to become better at it by allowing his crops to thrive.

Partnership seems a proper term. Some farmers wage war on these powers, attempting to assert the force of man and science on the world of nature, never knowing that their successes will be short-lived. Some other farmers more wisely seek partnership with that world.

The books tell the farmer that the health of his soil depends on the presence of billions of microorganisms, most of which we have not even identified. By adopting certain practices he can increase this population or, conversely, he can diminish it by ignoring common sense. The farmer gives thanks for being privy to the secret.

The books of an earlier day tell the farmer that his fortunes will rise or fall according to the ratio of earthworms, which are even better tillers of the soil than the huge green and yellow implements so common in farm country.

Where do these creatures come from? How do they get here? The farmer knows not, but they are here and he gives his thanks. And now that they are here, decidedly as partners, the farmer worries whenever he takes a tillage machine to the field. He fears disrupting their burrows and blocking the air channels they construct underground.

And what of the wondrous ladybugs? How do they know they are needed, how do they know when to arrive, and what is their mission? They have come in squadrons, purposefully patrolling the fields for black-hat bugs they snatch from plants under attack.

The farmer watches them and he learns. Again the books did not tell him this would happen, but the ladybugs for the first time worked their magic on the dreaded potato beetles this year and eliminated the need for spraying costly organic compounds.

These and other mystical occurrences give cause for celebration and the giving of thanks for being allowed to partake as a partner. These are things that drive the farmer to want to do it again next season and again the following season.

Beyond these gifts of nature, the farmer understands that he would be nowhere without the support and encouragement of friends new and old and even perfect strangers who came into his life and contributed in their own special ways.

Reflecting on it, the farmer is in awe. During the year, more than 30 city people, most of them working solely for the joy of it, came to assist in one way or another. They did the thankless work, the work that makes backs hurt and calf muscles sore. They worked without complaint. They brought laughter, made the farm a dearer place, helped make it succeed.

Oh, there were times, the farmer must now confess, when their work and their ways exasperated him and made him want to do his work alone. How selfish and small he was, the farmer now realizes after the frosts have slowed him down, to sully their joy and their caring.

They, too, all of them, are members of the special partnership of man and nature that the farm creates, drawn by the same forces that the farmer is blessed to experience each day of the year. He regrets they are not here this week for the sharing, but he honors them just the same.

It is the time of thanksgiving.

IV.
FRIENDS
AND NEIGHBORS

WHY ARE PEPPERS GREEN?

Oct. 3, 1990

There is a tomato patch, purposely planted in a low-lying place to collect moisture, that has bedeviled the farmer all year long. Healthy seedlings were transplanted in May, appropriately spaced in long rows that were appropriately separated, and then shrouded with a heavy straw mulch to hold in the damp and choke out the weeds.

For reasons unknown to the farmer, the tomato patch went berserk. The plants became huge, creating a thicket that covered the walkways and made passage nigh impossible. Billions of tomatoes seemed to form on the vines – large, well-shaped and, as it turned out, tasty.

The picking was difficult, but this patch was a success. Except for one thing. One of the four varieties of plum tomatoes in the field set almost no fruit. And to this day, the fruit that did form has failed to ripen.

It is a mystery for which the farmer has no explanation but for his partner's plausible theory that the intense heat of July probably fried the plants' flowers just as they would have begun forming fruit.

Puzzles of this sort arise routinely in the truckpatch. In fact, so confounding are the ways of nature that a day rarely passes without the farmer realizing how little he knows about his craft.

This benightedness, however, is not perceived by most of the customers at the urban markets who blitz the farmer with questions about his wares and about the problems in their own gardens. They imbue the farmer with undeserved omniscience.

Truth be told, however, the benightedness is not limited to the truckpatch. There was a time when just about every city dweller had a tie to the country – a relative, or whatever – that provided at least a cursory understanding of the source of our food.

People knew, for example, that a chicken planted upside down would not grow. They knew that milk did not come from a stockroom behind the dairy counter. They knew the difference between sweet corn (for people) and field corn (for animals).

Today, the city's ties to the country are more tenuous, and so apparently is the understanding.

One of the farmer's favorites was a woman who stood before his flower display at the Columbia market and quizzically eyed a bucket full of cockscomb, a large burgundy-colored flower that resembles an oversized brain.

She called the farmer over and asked, almost under her breath, "Is this a vegetable or what? How do you fix it?" When told that it was actually a flower that would dry and keep for months, she bought three stems and seemed pleased about it.

Another woman pondered the bunches of dried statice, made a choice, and moved tentatively to the counter. "Is this flower dead?" she asked, with a twinge of pain and hurt in her voice. It was a hard question to answer.

On a recent Sunday at Takoma Park, a serious young man picked up a red pepper and then leaned forward with a question. "Why is this so shiny?" he asked. "Because I buffed it all night," the farmer answered playfully, although in actual fact each of the peppers is wiped free of field dust before it goes to market.

A corollary question comes up repeatedly. A customer will survey the flower and vegetable display and then ask, "Why do your things look so colorful?" The farmer almost always responds that he uses a lot of spray paint, an answer that draws an understanding chuckle.

Vegetables, of course, come in many colors, and the farmer gets a perverse sort of pleasure from bringing in the offbeat things that are sure to provoke confusion and questions.

Just when the shopper learns that the Chioggia beet, with the bright red exterior, is not a radish, the farmer throws in a ringer. This fall, it's the Scarlet Ohno turnip, which gets confused with both a radish and the Chioggia beet.

Other items, such as the orange peppers, the burgundy snap beans, and the purple cauliflower, never fail to stir the curiosity. They are grown not for the fun of it but because they taste and they look good. Once tried, they usually are purchased again.

Another genre of questions is more challenging: "What does it taste like?" or "Which is the best tomato?" and "How hot is this pepper?" The farmer has a series of wise replies ("Like a Snickers bar, only cheaper" or "Sort of like a shag carpet" or "Acidic, but not too acidic"), but the best answer is to offer a sample.

Once sampled, the little Yellow Doll watermelon, with its canary-colored flesh, became a best seller. The squatty-looking Jenny Lind melon, once tried, became a favorite. The Sweet 100 cherry tomato

("Better than an M&M") was a box office hit. Ditto the Chioggia beet, with its concentric red and white circles and fine flavor when eaten raw. Old-time tomatoes, sometimes misshapen and ugly, were a summer sensation after tasting.

So there are fun and games in the truckpatch and the farmer, despite his inability to solve many of the mysteries that confront him each day, feels a certain usefulness when he can explain how to cook a kohlrabi or how to use the fennel and the frying peppers.

He also has learned to be more patient with the trying questioners, such as the gardeners who insist in the spring on buying a plant that will produce a red pepper instead of a green pepper.

It probably wasn't all that long ago, the farmer reminds himself, that he didn't know that most peppers start out green, then ripen to red, gold, orange, purple, or chocolate-brown. Another of those mysteries.

CRAMER THE CHEF HAS A FIELD DAY

Sept. 16, 1992

S ome time ago a friend from Iowa, where unending fields of corn and soybeans are grown to feed hogs and cattle, made a remark intended as facetious, but truer than he knew.

"How interesting...a farmer who actually grows food for people," said the friend, alluding to the fruits of the truckpatch that are hauled to farmers' markets and restaurants in the city.

Unwittingly, the friend had touched on one of the driving forces of the truckpatch – the pleasure and amazement the farmer derives from seeing how customers greet his produce and convert it to culinary delight with just a pinch of this and a dash of that.

The farmer, frankly, needs a certain amount of ego gratification and so to that end he proposed to his friend Cramer, the chef at City Cafe, that he come to the country again and confront the truckpatch at its lustiest, when summer fades to fall and the bounty piles high.

"David," the farmer said, "you can walk the fields, pick whatever you like and show us how you let the imagination take hold. No menu, no advance planning. Just show and tell."

Cramer could not reject the challenge. The work crew, nearly as intrigued as the farmer by the mysteries of the kitchen, awaited the visit eagerly. Peterson, the partner, readied the pots, pans, and whisks.

Cramer, undaunted by this gustatory Rubicon, came, saw, and conquered. He foraged across the truckpatch and through storage bins, snagging peppers, melons, potatoes, garlics, onions, fennel, celery, herbs, tomatoes, eggplants, and more.

The chef's eyes were bright. "Killer stuff, killer stuff," he kept muttering. He sounded like the same guy who spent three years as a sound mixer for the Grateful Dead. "You do the same thing with food," he explained. "Take what you've got, blend it, make it hum."

Into the kitchen he went. Chop, chop; whack, whack; stir, stir. The grill was readied. First came a rich dark marinade, into which went the boned chicken breasts. The chopping and stirring continued.

Jamaican jerk emerged, redolent with flavor and heat. Then came a melon salsa – antique Jenny Lind melons, mixed with peppers hot and sweet, onions, cucumber, herbs. Then nine kinds of cherry tomatoes went into a large bowl, bathed lightly in a vinaigrette. And then slabs of Italian eggplants, readied for the grill with oil and vinegar. Along with this, steamed rice (from some other farm).

The crew gathered in awe at the table. Mason, a worker inclined toward the kitchen himself, forked into the chicken and seemed struck dumb. "Extraordinary," he proclaimed. Cramer the chef glowed. And rising to the moment, Peterson trotted out a pie laden with truckpatch apples, raspberries, and blackberries. The pie vanished in a trice.

This was virtuoso stuff and this, the farmer thought, is why we do what we do, why we spend long hours on hand and knee in the fields, nurturing and cajoling the plants to produce provender that will become the joys of table that satisfy body and soul.

This, too, was Cramer's inspiration. Like a musician spurred by the applause, the chef was driven to an encore. "You guys up for another round?" he asked. The crew urged him on and Cramer came back a second day and resumed his tramping through the fields.

He gathered ingredients for the fiery Filipino adobo sauce that would go on his grilled steak and then he threw a challenge to Peterson. "Choose any vegetables you want and we'll see what happens," he said.

From this came a brilliant gratin of yellow Finn potatoes with leeks and fennel, thrown together in almost an instant; a side dish sauté of red onions and shiitake mushrooms; a seasonal salad of arugula, nasturtium leaves, baby kale, French sorrel and edible flowers. To accompany this, jalapeño pepper corn bread.

It was the fennel that especially seemed to lift the chef. As his big knife shaved thin circles from the creamy bulbs, he said, "Ah, fennel. It's the most underrated vegetable. They ought to just throw celery away." Well, harrumph, thought the farmer.

But whence the inspiration? the farmer wondered. "You go with what you've got," Cramer explained. "Seasonality regulates it all ... and that is what some of us in the restaurant kitchens are trying to get back to. We must rely more on our local farmers, push them to use their imagination so we in turn can use our imagination."

And with that, the crew again sat down to dinner. For a closer they got Peterson's homemade blackberry ice cream. At the end, applause rained down on Cramer's work of art. The chef basked in the adulation and seemed entirely ready to do a repeat the next day, but, alas, he had to move on.

It set the farmer to musing on the remark by the friend from Iowa. How interesting ... food grown for people. Seasonality. Freshness. Imagination. Creativity. Hard work and long hours in field and kitchen. Simple pleasures. "Extraordinary," as Mason the worker might have put it.

But then again, not really.

It's Quitting Time at the Farm

Aug. 8, 1990

On a slow day this spring, the touring manager of a Soviet state farm dropped in for a visit and the farmer was smitten by the instant rapport he found with his guest. Over coffee and cake, the young Russian detailed his problems, which seemed endless and overwhelming in comparison with those on this little American truckpatch, but the stories had a familiar ring.

Yes, he had 200 employees but few of them showed initiative. Yes, he had machinery but the workers abused it because they didn't own it. And when it broke, the machinery was left to rust in the fields for lack of parts or lack of mechanical talent. When quitting time came, the workers simply vanished whether or not the work was done.

"Our agriculture will never prosper until we have private ownership and more competition," the visitor said. "On our farm, we say the problem is that it is ours, but not mine...The workers do not have a vested interest, so they do not care."

What he was saying, in so many capitalist words, was that good help is hard to find these days – a cliché, no doubt, but one that the farmer grasped with ease.

Some of his city customers have the impression that the solitary farmer works his fields, harvests his crops, and takes it all to market as part of some virtuoso act of labor and ingenuity. Nothing could be farther from reality.

In the beginning, when the truckpatch wasn't much more than an oversized garden, life was simple. The farmer and his trusty accomplice, Peterson, divided the labors according to interest and ability, and worked until the chores were done.

But in time, it became too much. Too many weeds to pull, too many beans to pick, too many lettuce seedlings to transplant, and it was clear that help was needed. Thus came their introduction to the mystical world of personnel management.

Peterson and the farmer agreed on some basic points. The pay would be competitive or better; no worker would be asked to do a task that the farmers themselves would not do; ample breaks and rests would be insisted upon; the most onerous jobs would be reserved for the farmers; assignments would be varied constantly to avert boredom.

However reasonable that might appear on paper, the reality turned out to be different. A parade of workers – some excellent, some not – has come and gone through the truckpatch and the thought of a farm suddenly without a work force nags the farmer constantly.

A well-recommended young housewife in need of money came to work one day and performed capably and with promise. Then she vanished, sending word through a third party that since she lived four miles away, she could not afford the gas to continue here.

A strapping young man knocked on the door another day, begging for a job and bragging about his industriousness. He missed his first day of work because he "misunderstood" the starting time. On the second day, despite rigorous instruction in hoeing, he whacked off scores of valuable lettuces painstakingly transplanted by hand.

He missed his third day because of a pimple behind his ear that he feared to be cancerous. He missed his fifth day because his car had to be inspected. He missed his seventh day because he had to pick up his prom tuxedo. He couldn't work Fridays because they were reserved for his girlfriend. For one reason or another he missed a string of Saturdays and the farmer finally let him go.

A teenage girl worked all one summer and looked eagerly to weekend duty in the fall, a time when more hands are vital in the truckpatch. Her fall work plans were squelched by an unanticipated pregnancy.

Another longtime helper, an industrious and uncomplaining sort, had trouble with judgments, amounts, plant names, and details. When asked how many beans she picked, the answer was usually "a goodly amount." When told to pick spinach, she ravaged the Chinese cabbage. Plant names could not be remembered and thus came to be known as the "brainy" flower or "that purple flower" or some such.

Nearing terminal frustration, the farmer and his partner decided there had to be a better way. Like many of their truckpatching brethren, they would recruit outsiders – students, city people, ecology buffs – who surely were lusting for a bucolic summer on an organic farm and the redemption of good, hard work. Sure.

A nice bunkhouse was constructed over the winter to accommodate the workers who were certain to leap at the opportunities. An application form and a brochure were prepared and recruitment posters were sent to job-placement offices at 15 colleges and universities.

The response was not thunderous – only one applicant, an appealing candidate who ended up taking summer work with a seed company. Panic set in. Ads were placed in half a dozen university newspapers and many potential workers called back. Applications were sent to each caller. Women were interested, but not a single male caller applied for work after seeing the brochure.

The first hire, a budding sculptress, wrote rhapsodically about her concern for the environment and her desire to spend time in the country. She was coddled and looked after, to make her feel welcome, but she quit after a week, complaining that the life was too lonely.

Finally three other young women, two of whom had previous farm experience, were selected, and they have helped the truckpatch muddle through another season. Or through part of another season. They have announced they'll be leaving sooner than anticipated.

With that, and fearing a new labor crisis, the farmer decided to employ another job-seeking young acquaintance who had built his own greenhouse, planted his own garden, and had years of experience on his uncle's farm, operating sophisticated equipment.

He worked well on his first day, driving a tractor, hand-loading manure and spreading it on the vegetable beds. He hauled mulch and he plunged a lawn mower without complaint into a weed-infested strawberry patch. He helped build a new trailer. Here, obviously, was a gem of a worker.

That night, working alone in the cilantro bed, the farmer was upbraided by one of the female workers. "I'm very offended by the male chauvinistic attitudes around here," she began. "You hire this guy and he gets to do all these 'male' jobs – driving the tractor, building the trailer..."

"Wait a minute," said the farmer. "He's here because we're not getting the work done. He has experience and he has skills. Is it that you wanted to shovel chicken manure and spread it on the fields?"

"No, I didn't," she said. "I just don't like this chauvinism."

There was much palaver and explication after this and the farmer, although reeling from the jolt of the harangue, felt that a truce and an understanding had been reached.

After his auspicious start, the new young male employee never came back. He didn't call, didn't write, didn't even send an emissary to collect his pay. And the offended woman employee has never inquired about his disappearance.

The farmer sometimes feels like his Soviet counterpart. He tries, but he has much to learn.

GAMESMANSHIP
AT THE FARMERS' MARKET

June 13, 1990

Spring training, so to speak, was complete and the farmer was psyched up and ready for Opening Day, a ritual April event in which he envisioned himself being very impressive in his first turn at the plate.

Opening Day in this context was the first day at a farmers' market in the city, an event as important for renewing old friendships with customers as it was for again seeing farmer friends and stoking the old rivalries.

The farmer may make too much of this, but the race to have the earliest lettuce or the first ripe tomato is one of the things that drives him. It makes him take too many risks in cool weather, but the pride swells when he can best a rival from warmer Maryland or Virginia.

So the farmer this time picked his target — an expert grower named Francis Roland, who also is driven by the same desire to be first, even though he doesn't like to admit it. The farmer knew that Roland's goat would be gotten when lettuce from Dott, Pa., showed up before lettuce from Friendly, Md.

Roland took this humiliation with his customary good grace, but there was a glint of revenge in his eye. A couple weeks later, Roland tauntingly called to the farmer from across the market. As customers turned to follow the badinage, Roland held up one of the largest, prettiest, bright-green lettuces the farmer had ever seen.

"What in the world is that?" the farmer asked, now trying to hide his chagrin. "It's a Tango variety," Roland said. It took the farmer a while to realize that he'd just been put down twice — not only did he not recognize the variety, he'd just been taught that one does not harvest a lettuce before its time.

More comeuppances awaited the farmer in the ensuing weeks.

Sharon Blankenbaker showed up with a mountain of lovely radishes before the farmer had even put a radish seed in the ground. Susan Planck and Hana Newcomb humbled the farmer with their early spinach.

Andy Green came rushing up on a Sunday in early May, one hand behind his back and a smirk on his face. "I brought something for you," Green said, thereupon whipping out a field-ripened tomato. The farmer tried to make light of this tomato, noting that the fruit had been helped along by a greenhouse-like plastic covering, but he'd been stung and it didn't feel good.

The next Sunday Green went through the same drill, this time holding out a summer squash. Game, set, match. The farmer now was plainly on the defensive, left only to vow that somehow Green would be paid back next year.

All of this is in fun, of course, and it's mostly the extent of the socializing among truck farmers once their growing season begins. There is no time for visiting then, so the markets become their social center, a place to gossip, trade notes, and whine about the weather.

There are several dozen of these markets in the Washington area, which Susan Planck likes to call the most egalitarian of institutions. Where else, she once pointed out, can one see families of blacks, whites, Asians, and Hispanics, the young and the old and the handicapped all working together and caring about each other?

Well, probably no place. So there is grief when it is learned that Sharon has lost her husband and there is relief when a question about Roland's health is cleared up. There is pride when Jerry Worrell's daughter is chosen as queen of an agricultural fair and Green's daughter makes the track team at the Maryland School for the Blind.

The farmers feel good when they see a bright little first-grader like Jake Ricci, the son of Tony and Becky, making small bunches of flowers and offering them for sale to grade-school children at the market. They get pleasure from knowing that Nina Planck is doing well at college and that her brother Charles will marry in September.

This sense of family extends beyond the farmers to their customers, who greet their return to the city and inquire caringly about their crops and how they passed the winter. The farmers in turn learn the names of the children and remember the tastes of their parents, often setting aside coveted items for them.

All of this helps reaffirm the farmer's faith in the goodness of people, yet it still is something of a revelation to him. He is flattered that customers he barely knows invite him to their homes, that customers take the time to bring him recipes and baked goods, that a customer calls long distance to apologize for what she fears was an offensive remark at the market that morning.

And then there are the simply astounding things, such as the day that a woman offered Bill Waller, a personable chap who grows and sells plants, the free use of her condo in Montana if he followed through with his plans to vacation out there some day. Waller had met her only moments before.

But this is the way people are at the markets. Unkind words are not heard and denizens of the city, given the opportunity of face-to-face encounters, continue to show that they are American farmers' best and most supportive friends.

These are pleasures not known to farmers who grow the grains and drive the huge machines on the distant plains. These are pleasures of the truckpatch and they help to justify all the time spent on hands and knees, cosseting the carrot that will bring praise from an appreciative eater in the city.

It makes the race to have the first ripe tomato not so important after all.

GRATEFUL FOR ALL CREATURES GREAT AND SMALL

Nov. 15, 1989

A s the chill deepens and the outside work slows, the farmer turns to his ledger books, which are deceiving; they confirm that his financial rewards are small, but fail to reveal that he has acquired immense wealth.

Looking back over the year, the farmer realizes that he has become part of a family that somehow seems to count on him, that supports and encourages him and justifies the workaday drudgery, which in the truckpatch means much time on hands and knees.

In the formal sense, the farmer is not a man of religion. But the year has revealed new insights and so at this time of giving thanks, the farmer bows in considerable awe to the combination of forces of man and nature that have led him to his new wealth.

This has been a year of revelation, for the farmer has learned the folly of counting bottom lines in dollars and cents. How can he place a value on nature's teaching, which has helped and hindered his best efforts to husband soil and plants? How to price the proliferation of earthworms that have brought health to his soil? Or the invasion of protective barn swallows and ladybugs and praying mantises? Or the seductive aroma of fresh-turned soil?

Joy is derived from the observation of these processes, from the participation that helps speed them, and for this thanks are given.

All of this counts as part of the wealth and profit, just as does the acquisition of family. The people of this family, most of whose full names are not known to the farmer, are the customers who come to his markets to buy his produce and to share the intimacies of family and friendship.

Thanks are given to these people, especially the children, who seem to somehow know intuitively that life springs from the soil and that good things come to those who protect and nurture it.

A freckle-faced second-grader named Flannery, unhappily dependent on her mother for a ride to the market, comes early to help the farmer place signs around his stand. She has confided to her mother that perhaps if she does a good job the farmer will invite her to work on the farm next year. He will.

A toddler named Kelsey comes regularly in her father's arms. She cannot pronounce the farmer's name but she knows he will always place a ripe strawberry or some other delectable in her chubby hand. A shy 4-year-old named Victoria, who once drew a portrait of the farmer, averted her eyes all summer long and then suddenly embraced the farmer's kneecaps.

Jane and Amy, grade-school sisters, show without fail each Sunday. Their eyes light up in front of the vegetables, bespeaking a special appreciation. They brought pleasure to the farmer by arriving one day to "work" on the farm. The lettuce they helped cover with protective netting in the fall grew to marketable size, thanks to their effort.

The adults are just as supportive. A lady named Lani would arrive in the hot months with bottled ice water for the farmer. A man named Russell brought brownies, his wife Marci brought cake. A man named Phil brought news clippings, knowing that the farmer was out of touch. Judy brought brochures on gardening. A woman named Donna, a stranger to the farmer, dropped by to introduce Richard, her boyfriend.

An infrequent shopper at the stand came with antique seeds he wanted to share. A lady named Jill brought beer from Cincinnati and carrot seeds from France. Women whose names were unknown to the farmer brought recipes. A Redskins fan, a retired chap, routinely invited the farmer to stop by and watch football on his TV.

Others who did not bring gifts brought their care and best wishes. Each week the farmer would be peppered with questions about the bugs and the weather and sales. They worried aloud if the farmer would make it, offered their support, and showed that the distance between city and country is not as far as some policymakers would have us believe.

How does the farmer thank his family? How can the farmer respond to the more intimate details of life shared by these people who come for his vegetables and his berries?

One man, a regular shopper, began asking the farmer to pick out the best tomatoes and set them aside for him. "I hate to tell you," he finally said one day, "but I've just learned that I'm going blind and I can't see what you have on the stand."

Another elderly man, wobbly on half-crutches, always paid from a billfold wrapped in a plastic bag. "I've done this since I dropped the billfold in a puddle of rain ... couldn't get down to pick it up," he explained. Last seen, he was ebullient – about to go into a hospital to get new plastic knees, offering to come work on the farm next year.

Yet another part of the farmer's intangible wealth is the family of fellow vendors he has joined. They all compete, most of them grouse and gossip about each other's habits, but they care for each other. They support one another with barter and discounts and tips on growing their specialties. The farmer has learned from each of them and he thanks them.

He is indebted to a farmer named Francis, a shy man from Friendly, Md., who has taught him the meaning of gentle. He thanks Chip and Susan of Purcellville, Va., who supported the farmer by sharing their knowledge with grace. He thanks Sharon, who astonishingly refuses to bring her rutabagas to market so as not to offend the farmer who first introduced her to them. The farmer is grateful for the lessons taught by Tony and Becky, determined young Pennsylvania growers whose grit brought them back from the edge of failure.

This is the farmer's new family, his new wealth and in many ways the inspiration that allows him to ignore the aching elbows and tired arms and bruised knees that are the bane of the truckpatch.

He is in awe of these riches and he is thankful. Amen.

I GET BY WITH A LITTLE HELP

Nov. 24, 1991

The year has been long and difficult, readers of these dispatches from the truckpatch know by now, and there is little point in reciting the farmer's trials with drought, heat, and deer. In past years at this point the farmer has paused to give thanks for the abundance, for being allowed to glimpse the secrets of nature and for having good and supportive friends.

But this year, and this is said without exaggeration, the farmer is thankful as never before. Thankful that this season finally is over, thankful that he made it without requiring therapy, thankful that he can store 1991 as a bad memory and move on optimistically to next year.

Truth be told, there were times when the farmer felt he would not make it. As the drought deepened, the farmer's depression of spirit and energy intensified. As the deer and groundhogs rampaged through the fields, the farmer's fury and frustration towered.

And now that it's over, the farmer looks back and smiles just a tad about his hypersensitivity to events he could not control. He can in fact give thanks for new lessons learned about coping and for slowly coming to terms with his own fragility.

The additional truth that must be told, and for which thanks are hereby duly rendered, is that the farmer could not have survived had he not had the support of sympathetic customers and a magnificent crew of workers who in time came to be as dear as family.

An impression often is left, mistakenly, that the farmer works his magic in the fields all alone in communion with forces of nature that only he understands. This is not the case, for the success of the truckpatch is reliant on the care and skill of the hired hands.

The first to arrive was Paul, a quiet young man of quick wit who read philosophy and played John Lee Hooker tapes too loudly in the greenhouse. But the seedlings survived and who is to say it wasn't because they were nurtured by the blues?

Then there was Virginia, a natural-born steward who grasped the excitement of watching a seed push its first leaves up through the crusted soil. She asked serious questions that often were too difficult for the farmer, she studied books about the craft, and by the end of her tour she was offering solid suggestions about truckpatch improvements.

And James, a romantic who aspired to having his own farm one day and invested half a year in an apprenticeship that would neither dissuade him nor make him the more obsessive. The work was hard and often perplexing, but his triumph came in proving that he could stay the course. The farmer was never prouder.

And then Beverly, a mirthful soul who entered copious notes in a small theme book and was relentless as an interrogator. Imbued with an appropriate sense of the absurd, she will no doubt be a farmer someday.

Together they learned, sometimes quarreling, sometimes waning under the duress, but always cooperating and always cajoling, even when the farmer needed comeuppance. The success of the truckpatch was as much theirs as it was the farmer's.

A final truth is that none of this would have happened without the steady guiding hand of Cass, the partner known in these dispatches as Peterson, whose genius with plants and the ways of nature has a constant humbling effect on the farmer.

When the farmer was out of sorts, Peterson calmed the waters. When the farmer had no more ideas, Peterson solved the problem. When the farmer riled the crew, Peterson was a peacemaker. When plants rebelled at others' touch, they responded to Peterson's way. When the toll of drought seemed worst, it was Peterson who came up with the saving graces.

Sadly, because it is the way these things work, it is the farmer who too often gets the credit for successes of the truckpatch when it rightly resides elsewhere with others.

So yes, the farmer is thankful that it is over, this direst of dire seasons. But he is thankful, too, for Peterson and for the crew and for the loyal and supportive customers; for the chefs Perry and John and David, who encouraged and commiserated constantly.

Without them, there would have been nothing. The farmer thanks them for allowing him to ride along.

V.
THE BUSINESS
OF FARMING

DOWN COUNTRY ROADS
TO BIG CITY BUYERS

June 7, 1989

M ost of the day was spent in the fields, gathering lettuce, spinach, scallions, asparagus, broccoli, and other spring comestibles from the truckpatch and by nightfall the farmer was bushed. Each of these things had been carefully washed, bunched and boxed, and stashed in the cooler for loading onto the truck at dawn of the morrow to head into the market in the Washington suburbs.

Early Sunday morning, loaded to the gills, the truck headed down the interstate for the city. Thirty miles into the journey – too far to turn back – the farmer had a sudden dark realization. In the hubbub of preparation, he had forgotten to bring along his plastic food bags and shopping bags.

To paraphrase, the farmer muttered something like "rac-a-frasha." He may even have said it several times. He was disgusted with himself and now uptight, because trying to run a farmer's market stand without bags is akin to playing shortstop without a glove. It's very tough.

But in the end, it worked out.

Farmer-friends at the market came to the rescue with offers of bags. That was to be expected, actually, because even though the competition in these markets is avid, vendors are first of all sympathetic friends.

Of course, it could have been worse. One time the farmer blithely went to market, leaving behind a mountain of sugar snap peas that had taken hours to harvest. No friend could help on that one. Another time, the farmer failed to stash his broccoli in the cooler overnight. When the boxes were opened the next day, the farmer found that once-lovely heads had turned to flower in the heat. He muttered "rac-a-frasha" then, too.

These oversights are almost inevitable, for the day prior to market always is a day of dizzying activity. Too much to pick, not enough help, too much to remember, too little time in which to do it. Sometimes, when the bounty is at its fullest, the farmer must complete

93

the day's work with the help of the lights on his pickup or his tractor. The basil must go through.

To the tens of thousands of customers who flock to the markets in the Washington-Baltimore area, the sudden appearance of elaborate stands with fancy sun-screens and counters overflowing with farm-fresh produce may be viewed as just one of those things that magically happens.

In reality, it isn't that way at all. It happens because most of these small-scale capitalists, no matter their education levels or their experience, are consummate planners who soon learn that an overlooked detail can mean a disastrous day at market.

They learn to schedule their harvesting so the produce will be just right on the stand. They make long lists of things to do and they write reminders to themselves. When the pickings are slim, they come up with new things to sell – ravage that lilac bush for fresh flowers, harvest that dried pigweed and peddle it as a decorative "amaranthus."

They learn that packing a truck is an art. They read newspapers to divine the price trends; they read the food pages to find out what titillates the cooks these days. They learn to improve their lettering, the better to make appealing signs. Some go so far as to study the path of the sun to pick their place in the market – a step aimed at protecting tender veggies from the hot glare.

The business of market farming is entirely unlike anything else in agriculture. The grain farmer, for example, hauls his harvest to a country elevator, offloads it, and waits for the check. The livestock producer sends his animals to market when they're ready and takes the going price, for an animal ready for slaughter can't be held back.

The truck farmer's craft is different – something simply not taught by our schools of agronomy. He (or she, for there are many female truck farmers in this mid-Atlantic region) learns by serving a sort of apprenticeship, watching others and innovating where he can to capture a niche that no one else has exploited.

A baseball analogy is in order. Just as most big league ball players spend learning time in the minor leagues, so do most truck farmers. Some begin by selling their produce off the tailgate of the pickup; others may start with a small counter under a shade tree outside the farmhouse. Thus begins the learning process.

In the case of this farmer, the process began at a small market in Hagerstown. His first day there, which now seems like years ago, was full of angst and uncertainty. His spinach, neatly bunched and easily the best in the market, didn't sell. He was stricken. Weeks later, he realized that Hagerstown customers were accustomed to buying their spinach already bagged. First lesson learned.

His first week's take was $25. Poor, he thought, but it was a start. The second week's take was $11. The third week fell to $10. This was not the beginning of a brilliant career in direct marketing or in feeding the millions as a noble steward of the land. Various permutations of rac-a-frasha were heard.

When the farmer's despair became palpable, two wise and generous fellow marketers, Lois and Richard Oberholzer, counseled patience. Richard always insisted "there is no money to be made in vegetables, at least around here," but the Oberholzers had persisted and customers regularly queued up at their stand.

Things will improve, the Oberholzers said repeatedly, and in fact they slowly did. The customers slowly began to come, looking for the different or the unusual, often returning with praise for a carrot or a green bean, and the farmer's spirits lifted. He knew then why Lois and Richard had persisted: there was no greater joy than pleasing a carrot lover.

But a day came when the farmer had served his time in the minors and he had to move on to the Big City, where there were more customers and more money, fewer home gardeners and wider disillusionment with the cardboard produce of the supermarket.

The farmer knew he had arrived when the customers would line up at his city stand, sometimes 15 minutes before the opening bell, just to pick out the tomatoes or the broccoli they would eagerly buy. It was like hitting a line drive into the seats at Memorial Stadium.

But the joys are often short-lived. The farmer still forgets his plastic bags. "Rac-a-frasha."

BROWSING FOR THE WISDOM
OF THE AGES

Jan. 23, 1991

C onfined to quarters by winter cold, the farmer on a recent afternoon was perusing the minutes of the 1880 meeting of the state
fruit growers' society when a thought occurred. In agriculture, the
farmer was reminded again by this antique volume, there isn't all that
much that's new.

There was a lengthy discussion about whether birds were really
the fruit grower's friends. There was talk about the use of soft coal
ashes for insect control in the orchards. A committee was appointed to
study the use of lime for fungus prevention.

Bird problems, insect problems, fungus problems continue to
plague modern-day orchardists, despite the arsenal of chemical solutions now available, and committees are being named all the time to
study these venerable intrusions.

The other thing that struck the farmer from his reading was that
the society, even 110 years ago, was meeting in January. Logical, of
course, because winter is the slow time for fruit and vegetable growers,
who are either looking back or looking forward.

Just as they were going to meetings a century ago, farmers continue to spend their winters in what sometimes seems to be a never-end-
ing succession of seminars, conventions, forums, and conferences.

There is some value in these convocations, although more often
than not they serve as an excuse for socializing or as a vehicle for academicians to expound on matters that tend to glaze over the eyes.
Politicians find the meetings useful, as it gives them another forum for
extolling the romantic virtues of the family farmer.

Modern farmers have developed these meetings into such an art
form that their 19th-century brethren would likely be astonished.

Mimicking doctors and lawyers and teachers, the major farm
groups hold their winter conventions in exotic warm places that presumably make it easier for talking turkey and confirming their own
notions of just how bad things really are back on the farm.

The farm magazines have refined this even more. They regularly advertise winter "educational" tours to Australia, New Zealand, Japan, China – almost everywhere, in fact – where farms are visited and costs are written off as a business expense.

This farmer is not immune to the virus of winter meetings, although he has neither the temerity nor the resources to hie himself away to some foreign land to study the intricacies of growing radicchio or parsnips.

This winter's travel agenda includes meetings in New Jersey, West Virginia, Maryland, Virginia, Pennsylvania, and the nation's capital. Largely because of guilt about leaving the farm or a fear of terminal boredom, the farmer has rejected other travel opportunities.

There is much to be learned, however. One conference features a talk by an Ohio farmer who teaches how to sell a $2 pumpkin for $10. Another offers a day-long course in soil management for a mere $380 tuition fee. The sponsor also happens to be a fertilizer company.

Another conference includes tours of innovative farmers' markets, which seems a curious thing to do in the dead of winter when not so much as a leaf of spinach is grown. Still another lures Eastern farmers to California to hear speeches by Eastern farming experts.

So the farmer, if he is worth his salt, must attend at least one or two of these events just to remind himself why he doesn't go to more of them. If it's socializing he wants, the farmer can go down to the general store at Dott and talk all day.

Rather than gallivanting around to these faraway winter meetings, the farmer finds it more useful to stay at home with the 1880 minutes of the fruit growers' society and other antique readings that tend to teach and provoke the imagination more than any of these public gatherings.

It was pure serendipity, but in the 1880 minutes the farmer came upon more important clues in his search for a low-cost, safe, and effective way to control pests in his vegetable plots. The answer, obliterated by modern chemistry, lies in these old-time books, and so the farmer's reading intensifies.

Those old minutes, by the way, reflected intense debate over the problems caused by birds in the orchards. "Let us see where we stand," said one Josiah Hoopes. "Mr. Merceron kills all the robins, Mr. Sprout all the pheasants, and Judge Stitzel all the sparrows. I have never seen the green worm in the city since the sparrow has been introduced. In England, we see [sparrows] by the millions and they are considered useful."

The meeting ended, reassuringly, with one member surmising that the sentiment of the society was in favor of birds. The farmer felt good about that, a little smug even, for he had learned seasons ago that his best friends were the birds that he coaxed to the farm to help patrol for bugs.

And he probably learned that from an old book on a cold day in January.

A Black, Blue & Green Thumb

Feb. 21, 1990

The thought probably lurked deep in his mind anyway, but it wasn't until the farmer fiercely clobbered his left thumb with a hammer, trying to drive a nail, that he finally made the connection between a piece of lumber and a sprightly head of lettuce.

The lettuce, you see, would grow from a seed planted in this greenhouse the farmer was trying to erect. But if he couldn't drive his nails straight and square the cuts of his lumber, the greenhouse would be unstable and the lettuce seedling might never reach the field.

Thus, the farmer began to fully realize what other farmers have known for centuries. In order to survive, the farmer must have skills that go well beyond knowing the soil and all the tricks of nurturing a plant to lush maturity.

He must be, among other things, carpenter, plumber, electrician, roofer, engineer, architect, engine mechanic, cement finisher, and drainage designer to make the farm run smoothly and to provide all the accouterments required by his lettuce seedling.

It can be reported that the new greenhouse, after many clobbered thumbs and mis-cut 2-by-4s, rose to completion in time to receive the first seeds of early lettuce, broccoli, and cabbage. It may turn out to be airtight and possibly stable enough to endure fierce wind.

Earlier, when temperatures were deep down in the teens, the farmer and his partner studied their how-to books and laid down a cement pad. The pad was reasonably level and smooth. On this pad they erected a large box that would become the walk-in cooler that would protect the mature lettuce before it was sent off to the market in the city.

Acquaintances assured the farmer that wiring the box for electricity was a cinch, nothing an amateur could not do. The farmer returned to the how-to books, but finally gave up. Too complex and too dangerous, he concluded. His friend Finn came and hooked up the wires with a self-assurance that was humbling.

So the cooler was finished and visitors praised the job. But the farmer felt embarrassed. The corners of the box were not square, the sides did not meet flush, the ceiling had a slight warp, and somehow the insulation sheets had not been cut straight. The professional who was called to hook up the compressor was kind enough to withhold comment on the flaws.

By now, the farmer had become learned and realistic. For his next winter project, a little cabin to house some of his summer help, the farmer understood that he needed help. His books explained the steps, but he still had no idea how to lay a foundation, no idea about load-bearing walls or notching rafters or sloping a porch.

That is when a man named DeLeeuw came into his life from the next county. DeLeeuw was known to the farmer for the fine lettuce and the exquisite flowers that he grew. DeLeeuw was known to him for the care with which he marketed delights from his own truckpatch.

What the farmer did not know was that DeLeeuw was the consummate agrarian – expert, it seemed, in everything he did. He could roof a cabin with the same ease that he made fresh chevré from his goats' milk. He could square a corner as adroitly as he grew fancy beets. He had as little trouble plumbing and wiring as he had with his leeks.

So DeLeeuw became the farmer's teacher. With a patience and care unknown to the farmer, he spent hours with his tapes making certain the cabin foundation was perfectly square. By example, he showed the farmer how to cut a microscopically straight line and how to align the walls of the little building. DeLeeuw clucked with gentle disapproval when the farmer sank a nail off center or mis-measured a rafter.

DeLeeuw scrambled up and down his ladders, furiously hammering and measuring and plumbing his lines. He showed the farmer how to carry a heavy plywood sheet up a ladder with little exertion. He showed how to make the weight of a long beam do the carpenter's lifting.

These were great new revelations to the farmer, who secretly wished he could take these teachings and go back and rebuild the cooler, go back and reshape the greenhouse corners to perfect square.

On the night DeLeeuw nailed the last roof beams into place, a great glowing moon shone in the sky and illuminated the building site. "It's so bright out here we could keep on working until midnight," the carpenter said. "We could finish nailing down the roof tonight."

But he stopped his nailing then and scurried around a nearby pile of brush. He hacked the top branch from a fallen mulberry tree and climbed back to the roof, where he tacked up the limb to symbolize the topping off of his beautiful little cabin.

DeLeeuw then called for a bottle of beer and sat with the farmer in the half-light, modestly admiring his handiwork and speculating on the pleasure that some future worker might find on the cabin porch with a summer moon beaming down.

There on DeLeeuw's porch, in the unseasonable balm of a February night, the connection between the perfectly cut 2-by-4 and the perfectly grown head of lettuce made itself crystal clear.

The farmer's left thumb still hurt mightily and the pain of his carpenter's elbow persisted, but he thought the lettuce might be the better for it this year.

REAPING A BUMPER CROP
OF PAPERWORK

Sept. 4, 1991

There was a time, still recent enough that the farmer can remember it with fondness, when his life in the country was kind of an idyll, free of the stresses of the city and far removed from the paper chase that threatens to engulf mankind.

Urban friends envied him and coveted these presumed liberties. The farmer himself felt a satisfied smugness about his good fortune. Free, free, free at last.

Then something strange happened. The farmer awoke one day to find that he was not free at all. Despite his best efforts to focus solely on his tomatoes and lettuce, he had in fact become a slave to distant computers and document shufflers.

The farmer was reminded of this when newspapers hereabouts reported excitedly that federal drought disaster aid had been approved for this and other counties burned by the summer blast. Just fill out the applications – paper, paper, paper – and be saved.

Such stories may reassure sympathetic city folk who think kindly of their food producers, but it strikes the farmer as not much more than pretty political theater.

Some drought victims, whose finances are twisted enough, will have the privilege of acquiring additional debt through low-interest loans. Some others will be allowed to make hay on their government set-aside fields – not a big deal when it was scorched as badly as other fields.

Most truckpatch farmers don't even think about applying. Stories are legion about the inability of the government inspectors to grasp the value of a burned rapini crop, blowtorched sugar snap peas, or a lost planting of strawflowers. The hassle isn't worth the help.

The bottom line is that the farmer has become acutely aware that he must at all costs avoid plunging himself any deeper into the paper chase. Wherever he seems to look these days, a computer is spitting printouts toward him and someone is sending him papers to fill out. Much more of this and there'll be no time left for the tomatoes.

To sell his food in the city, the farmer must fill out lengthy applications for a place in the farmers' markets. Then he must have his county agent attest that he grows only those items that he sells. More paper, more phone calls, more trips to the county seat.

To stay right with the law, the farmer must apply for an employer identification number and register with Internal Revenue. He then must withhold income tax and social security from his workers' checks. More paper, much more time.

To protect himself and his field hands, the farmer must apply for expensive workmen's compensation insurance. The farmer corresponds on a quarterly basis with a computer at the insurance office, which never seems quite certain how much the farmer owes in premiums.

To get his seeds, the farmer mails his orders to more than a dozen companies each year. The seeds come back in dribs and drabs, each shipment containing a computer note explaining why this or that item wasn't available and when it might be shipped.

The farmer must keep track of each of these mailings, lest his prepaid special eggplant seeds get overlooked in the shuffle. More paper, more time. Fearful of more runaround, the farmer simply wrote off the failure of two firms to send back-ordered lettuce this year.

To obtain parts for his tractor and implements, the farmer orders by long-distance telephone and then hopes that the parts depot got it right. Wrong parts came twice this summer and a disabled tilling machine sat idle until a computer in Detroit selected the right items. More paper, more phone calls, more time.

To maintain the organic certification of the truckpatch, the farmer each year must fill out an interminable document that reviews the history of every field and the soil improvement and pest control techniques employed. More time, more paper....

To pay for the privilege of the certification, the farmer must keep records of his sales and fill out quarterly forms detailing the fees he must pay for using the certification seal.

To make certain his fields are tended, the farmer reviews the job applications that come by mail and telephone, checks the appropriate references, and decides whom to hire. Letters and application forms are exchanged. Phone messages are left. Time, paper, time.

To get the attention of his state game commission, whose deer are trampling over his crops, the farmer makes repeated phone calls and writes letters of complaint that seem to vanish into a black hole. More time, more paper, more frustration.

And to write these letters, the farmer uses a word processor that occasionally will not work. He calls the computer company and tries to work out his problem with technicians who seem to know no more about the machine than the farmer knows.

There is more, but the point is made. It wasn't supposed to be this way, and in fact it did not used to be this complicated. But then without all this he couldn't be a farmer. And without it there'd be little to complain about. Which would be very unfarmerly.

A FARMER CAN'T BE LISTLESS

=⯈·◉·◀⯇=

Jan. 19, 1992

Getting ready for a new season in the truckpatch, from the farmer's point of view, probably is not unlike the preparations made by a Joe Gibbs before he launches into another year of football. Before the first kickoff can occur, or in this case before the first serious work begins in the field, much must be done to assure that all the pieces and all the players are in place.

Most of the seeds for his new growing season have arrived, some already have been sown in the greenhouse, and the farmer's enthusiasm over the new year intensifies.

But the farmer is confronted every morning by a long yellow legal pad that sits on the kitchen table and it tells him, even screams at him, that much remains to be done before winter is officially declared over and gone here on the farm.

The yellow pad is what the farmer calls his "to-do" list and it is his road map for every day of the year. The list changes constantly as one task is completed and another new one added. But there is more to the list than its practicality.

The to-do list is a psychological crutch of sorts. The farmer and his helpers get a refreshing lift when they can wield a red crayon and mark off a task completed. They feel a downdraft of new pressure when another job is added.

In fact, it seems, the list is never ending. It also is daunting in some ways, for it implies invoking many skills, some of which are well beyond the farmer's ken, some of which have been acquired by dint of need, and others of which never will be mastered.

Knowing how to plant a seed, nurture it, harvest it, and take the finished product to market is only a part of the demands of the truckpatch. The to-do list requires the farmer to be mechanic, carpenter, supply specialist, labor negotiator and employer, writer and graphic designer, plumber, banker, and a host of other arts.

Oh, it would be simpler to call in an expert and have the job done right. But there is neither money nor time enough for the farmer to rely on others. When a tractor goes dead, he can't wait on help from

105

afar. When wind rips the cover off a greenhouse, it must be replaced immediately.

So the to-do list is the farmer's master. It determines how his time is allocated and in its cryptic code language reminds him every day that the work on the farm is never done if the place is to flourish and succeed.

On a recent morning, the to-do list was sobering. Land of Goshen, the farmer exclaimed to himself, what have you done all winter?

This is what the list told the farmer:

❏ Pull V3 HT – This means disconnecting, rolling on a spool, and storing the hot tape – an electrified nylon strand – that is intended to keep deer away from a six-acre patch known as Veggie 3. Estimated time for the job: half a day.

❏ V3 Stakes – This means uprooting the fiberglass posts that hold the tape in place. Clearance needed for tractor-tilling of Veggie 3. Time: half a day

❏ Remulch Strawbs – Because of wind and pawing by deer, a straw cover must be replaced over two beds of strawberries, each 300 feet long, to protect tiny buds from cold. Time: most of a day.

❏ Prune Heritage – To clear the way for this year's growth, four beds of Heritage raspberry canes – rows of canes totaling about half a mile – must be pruned and the debris removed and burned. Time: a day.

❏ Prune Chesters – All of last year's fruit-bearing canes must be cut from the Chester blackberry patch and hauled to the burn pile. Remaining canes must be tied to the trellis. Time: a day.

❏ Tune FarmAll – A 50-year-old utility tractor must be tuned up, filters changed, and scratches retouched with red paint. But parts must be ordered first. Time: a day.

❏ Paint Equipment – A rotary mower, badly scarred by rust, needs to be sanded, repainted, and lubricated. The rotovator, a machine that tills the seed beds, needs the same treatment, plus new tines. Time: two days, maybe three.

❏ Cold Frames – Torn by wind and cats' claws, plastic covers on the wooden cold frames, where plants are set to harden off, need to be replaced. Time: half a day.

❏ Sanitize Cooler – Before it can be used again, the walk-in cooler must be emptied, swept and then washed down with a bleach solution to eliminate germs and dirt. Time: half a day.

- ❏ Finish Trays – Most of the job is done, but several hundred more greenhouse flats must be scrubbed and sanitized with bleach before they can be returned to duty. Time: most of a day.
- ❏ Van Check – The farm's produce van needs to go to the shop for its regular physical and a cracked wheel must be replaced. Time: too long.
- ❏ Get Benches – More metal table frames, concocted from scrap at a nearby tractor dealer, must be obtained. Easy job, but farmer must also schmooze with the dealer. Time: half a day.
- ❏ Redo GH Boxes – Large boxes used for growing some herbs and early market produce in a greenhouse must be refitted – plastic liners put in place, special soil mixed and poured in. Time: two days.
- ❏ Inventory Boxes – Vegetable cartons, tomato and berry flats, berry boxes, picking baskets, twist ties, food bags, and shopping bags must be inventoried and new supplies ordered after comparing dealers' prices. Time: a day. Pickup of same from various dealers: another two days.
- ❏ Irrigation Stuff – Hoses, drip irrigation tapes, couplers, cutoffs, clamps, and other watering supplies must be checked and then reordered and picked up. Time: two days.
- ❏ Potting Mix – Needs will be calculated and then about 50 bales of a growing medium – two tons, approximately – used for greenhouse starts must be ordered and picked up from a supplier three hours distant. Time: a day.
- ❏ Seeds – Seeds of some flowers and vegetables that turned out to be unavailable when ordered last month must be found in other catalogues and ordered soon. Time: half a day.
- ❏ Pallets – More pallets are needed for use in the greenhouses and around the farm. Time for hauling: half a day.
- ❏ Recipe Sheets – Recipes must be selected for printed sheets that will be given to farmers' market customers as a goodwill gesture and sales enhancer. Time: three-four days.
- ❏ See Printer – Meeting needed with printer in county seat to discuss production and cost of recipe sheets. Time: a day.
- ❏ Build Shed – More storage space to keep implements away from inclement weather is needed urgently. Lumber needs must be figured, supplies purchased, and structure erected. Time: four days.
- ❏ Prop Up Shed – Wind has blown a storage shed off its stone piers. Building must be propped up, piers fortified. Otherwise, the building is lost. Time: half a day.

❑ Stakes, Wire – To keep deer away from very sensitive greens patches, new fencing must be erected. Estimating needs, finding right supplies, setting posts, and attaching wire will take several days.

❑ Barn Loft – Construction of a barn loft for drying and storing flowers is urgent. Lumber and supply needs must be estimated and purchased and loft must be built (hopefully with help). Time: five days, maybe more.

❑ Shower Stall – To make life more comfortable for the work crew, a second shower – a solar-heated outdoor affair – will be built. Need to finish design, calculate supply needs, and construct. Time: two days.

❑ Turn, Order Manure – When weather allows, the composting pile of chicken manure used for fertilizer must be turned with the tractor. A new supply of manure must be found and purchased. Time: a day.

❑ Asparagus Beds – As soon as weather allows, asparagus beds need to be reconditioned with tiller, then fertilized and straw-mulched before new growth begins. Time: a day.

❑ Get Straw – Calls must go to local farmers to find a supply of straw or hay for mulching. Supplies will be short due to drought, prices will be higher, haggling will be intense. Time: half a day.

The good news is that the farmer has a clear picture of what he faces in the coming days. The bad news is that the to-do list just keeps on growing.

PLOWING THROUGH DINNER

July 10, 1991

B etween the drought, which is ruining half of the crops, and the deer, which are stealing what remains, the year is unfolding as something better left untalked about. If serious rainfall arrives soon, allowing the fall crops to be planted, the disaster may be mitigated. If not ... well, the farmer is left to ponder what might have been.

What might have been, among other things, is an awesome array of vegetable varieties and colors and flavors that parade across the truckpatch in an ordinary year. Just knowing it's there – whether it's eaten or not – is one of the farm's secret little joys.

The farmer's friends and acquaintances assume, of course, that his table groans from the weight of the produce and that the farmer's daily fare is a thing to be envied. Sad to say, that's not quite the case.

The truth is that the farmer doesn't eat as well as he ought to because when the fields are lush, there is never time enough to prepare the special dishes that ought to be de rigueur. Worse yet, when the abundance is greatest, there's no time to can or freeze foods for the winter.

After a long summer day in the fields, it often is easier to pop a frozen pizza into the oven for supper or to throw a hamburger on the grill and then adorn the plate with potato chips and factory pickles.

This leads to the additionally appalling fact that the farmer's tuna casserole in January contains store-bought canned peas and his spaghetti sauce in February often is made with tomatoes grown somewhere else. And, horror of horrors, his winter salad is often from a supermarket.

It is that way on many farms these days because life has become too complicated to put things aside for the lean times. There are farms in these parts, in fact, that don't even have gardens because the families are too busy, too lazy, or too indifferent.

During the last big drought, a neighboring cattle and grain grower talked at length with the farmer about planting his own "survival patch" – an acre or so that he envisioned seeding with basic vegetables to feed his family and relatives.

Predictably, that survival patch never materialized. The cattleman took the nicely prepared plot of ground that was to have fed his family and planted it with alfalfa the next spring. And several years later it is still alfalfa, feeding cows instead of people.

Now eating habits may not be good in the truckpatch, but they're not that bad either, because life is not all pizzas and burgers.

The farmer's diet has become attuned to the changes of the season and the ongoing abundance of the truckpatch actually has spoiled him to the point of avoiding supermarket produce whenever possible.

That is to say, when it is time for turnips and rutabagas, the farmer eats them with gusto and keeps seeking new ways to prepare them if he has the time. When squash, peppers, and eggplant are thriving, the farmer hies off to the kitchen to make a ratatouille that will last for days.

When the lettuces and the greens are lush, the farm is a mammoth salad bar. The first vine-ripened tomatoes are eaten voraciously for weeks and then abandoned in favor of whatever might be coming in next.

But the best eating here occurs right in the fields in what might be called "grazing." A sun-warmed tomato eaten just as it is plucked from the vine has no equal. Ditto, the sugar snap peas and the berries. With only a pocket knife for his utensil, the farmer has made many a full meal of his field-ready melons.

It is a process that begins when the first asparagus is ready – a just-picked spear, crisp and still oozing moisture, cannot be improved upon by the finest chef. Then on to radishes, scallions, peas, and snap beans as the season progresses. Then the cherry tomatoes and sweet peppers and ... on it goes until well past the first frost.

Even the grazing process is regulated by certain rules. The farmer eats the "seconds" – scarred or overripe items that would not do well at market. Seconds and rejects go into a "free-for-all" box in the cooler for the farmer and his helpers. Some other things, raspberries and strawberries, for example, are just too valuable and too scarce to be eaten in large quantities on the farm.

The concept is difficult to understand for the worker, who more often than not is a city person drawn to the farm by the attraction of a cornucopia of fresh veggies just here for the taking. Strictures about eating the profits become even more acute in times of drought, such as now.

There was, for example, a certain tomato plant that was watched covetously by all hands last month as it moved toward offering up the first large, red-ripe fruit of the season. At the precise moment of the tomato's readiness, a worker approached the door at lunchtime.

"There's a tomato out there that has my name on it," he announced. "Can I have it for lunch?"

The heresy was rejected out of hand. There would be other tomatoes later on, but that tomato had someone else's name on it. It was, in fact, too valuable to be eaten on the farm. Hard-hearted and cruel, perhaps, but only the farmer in a drought would understand.

SHH! IT'S CROP SECRET

Oct. 14, 1991

O n a recent evening, as is increasingly his wont when the season seems to get too long, the farmer was mumbling to himself. The mumbling grew, seemingly headed toward becoming another of those autumnal whines that farmers everywhere engage in to subvert blasé city folks' belief that all is hunky-dory out in the country.

The old pity-me-I'm-a-farmer syndrome, you might say.

Well, in fact, the tomatoes and the peppers had been lousy – too much moisture and not enough heat diluted the taste. The melons were bland, the cut flowers pooped out early, there was frost in late May, severe hail mowed everything down in September.

And so the litany went, about to persuade the farmer that he had every reason in the world to be morose, when Peterson emerged from the kitchen bearing a steamy plate of globe artichokes and a supply of drawn butter.

"What's with all this mumbling?" she asked. "How can you whine about a season that gives you artichokes? You're not even supposed to think about growing them around here."

These artichokes were superb – meaty, tender leaves that melted in the mouth – and it was decided then and there that this experimental planting would be expanded next year to marketable quantities.

The success of the artichoke planting, which was Peterson's idea in the first place, was another reminder that no season in the truckpatch is ever as bad as it seems or as bad as the farmer wants it to seem. There's always something new that does spectacularly well.

An underlying idea is to keep testing new things in an effort to outdo the competition at the farmers' markets. Attract more customers with better-tasting and better-looking vegetables varieties and the people will buy them and come back for more, the theory goes.

Enough other farmers scoff at this approach that they disqualify themselves as competitive threats. But enough others, because farmers are inveterate copycats, pay attention and emulate. And as they succeed with a borrowed idea, the truckpatch tries something else just to keep a leg up.

This kind of competition is so intense in places in California that some growers become manic paranoids. One of the farmer's friends out there, tired of having his ideas copied, will no longer talk publicly about the varieties he grows. He has barred other farmers from visiting his place and he prohibits cameras on the farm.

Things are not yet that intense in these parts, but they will be, and the farmer sees himself becoming increasingly circumspect about sharing his information. For example, eight kinds of radicchio are being grown in the truckpatch this fall and Peterson and the farmer know which ones are working best, although their lips are sealed.

As another example, the quest for a better eggplant continues in the truckpatch. There are black ones and green ones and lavender ones and pink ones and purple ones and another with striping that looks like a bowling ball. Customers scooped these up by the boxful while the farmer's friends return home with their unsold traditional eggplant. There is a message here.

The same searching turned up two varieties of chard from Italy that outperform the chard offered in the American catalogues. The searching came up with two kinds of kale from abroad that are sweeter and statelier than their American counterparts. They, too, will become regulars in the truckpatch.

Ten varieties of potatoes, half of them new to the truckpatch, were grown this season and their acceptance exceeded the farmer's fondest hopes. The first to sell out was an all-purple variety with a remarkable taste and texture that the farmer originally resisted because he thought it too much of a novelty. Wrong again.

Peterson, who is even more obsessive than the farmer about these new things, had a right to gloat about her successes. The cardoon that she envisioned as a great vegetable turned out to be even better as a variety of drying flower. Her mysterious old-time celtuce grew very well and made more sense after friends passed along recipes.

The French horticultural bean that she finally grew after talking about it for years made her seem visionary. It was, it turned out, a bean known to many, but rarely seen in the produce section nowadays; a bean that stirred distant memories. Customers called it the cranberry bean, the bird's egg bean, the October bean, and the borlotti bean. By any name it was a success and earned a place here next year.

So what started off as mumbling that wanted to become a piteous whine was really out of place. The tomatoes and the peppers might have been a joke – and they were – but who cares? The Bay City cantaloupes, grown from two Michigan friends' seeds, were very nice and the habanero chili peppers finally showed the meaning of h-o-t.

The new tomato from Italy and the old one from Greece outperformed their American counterparts, and the golden turnip purred with creamy sweetness. And the Japanese purple mustard ... well, all these things and more assured themselves space in the new season's plantings.

So no mumbling, no whining, no moroseness. As much as it might run counter to his inclination, the farmer is forced to admit that it has been a pretty good year after all. Mumble, mumble.

BEHIND THE LINES
IN THE TOMATO WAR

***

Aug. 9, 1989

Enough has been said and written to fill volumes about the plastification of the great American tomato, but one more new word of lament must be voiced. The supermarket tomato, as it is commonly known and despised, is a flat-out disaster despite the efforts of the grocers, the big-time growers and, yes, even the Department of Agriculture to put the best face on it.

Bred to look nice, ship well, and live long without refrigeration, the contemporary tomato has become a parody of itself – inedible, over-priced, and nothing like the delectable fruit that sends true tomato appreciators into ecstasies.

Even now, at the height of the local growing season, when one might expect to find an adequate tomato at the grocer's, the quest is fruitless. The tomato remains hard-skinned, pasty, devoid of taste.

To further the insult, they call it "vine-ripened," which in the trade actually means that it was picked green when it was at the first point of turning color. This is no substitute for the natural maturation that occurs when the tomato is left on the vine to grow to its peak of flavor and texture.

If one accepts the USDA assertion that four out of five Americans consider the tomato their favorite homegrown food, it is little wonder that millions of taste-conscious consumers try to grow them on patios and in backyard gardens. They are in search of the real tomato.

One alternative for the tomato lover is the farmers' market, of which there are more than a dozen of good size in metropolitan Washington. In these markets, the small-scale truck farmer who is motivated by high quality and consumer approval may be the last great hope for the discerning tomato buyer.

And therein lies the farmer's lament. In one sad case after another, the farmer observes that his friends and competitors inexplicably are falling into the habit of growing lousy tomatoes – the selfsame varieties that show up in the supermarkets.

When there are literally dozens of varieties available that provide the tomato taste and texture that tantalize, more and more small-scale growers seem to be turning to the commercial types that are the crux of the problem in the first place.

"Some of the tomatoes offered today are just not the type of varieties that we should be growing because they don't eat well," says Clark Nicklow, a veteran tomato breeder at the University of Massachusetts Experiment Station.

Nicklow is part of a dwindling band of plant scientists who put a high premium on flavor. For his part, Nicklow tastes thousands of tomatoes every summer, looking for the type that suits his palate. His tomatoes, both commercial and home-garden types, have won prizes for their taste characteristics.

"There is tremendous pressure on the seed companies to come up with tomatoes that do well, so they ignore the eating quality. There is nothing on the market that comes up to my standards," he says.

Why, then, would the small-scale farmers whose fortunes are tied so closely to consumer acceptance fall into the same trap as their bigger brethren? The answer is elusive.

They grow these plastic-like varieties that Nicklow so deplores because they presumably are seeking the uniform size, higher yields, and greater disease resistance that many of the hottest new commercial types offer.

Then to compound the misdeed, they pick their tomatoes well before their time, as they are just entering the orange stage of maturation. They often do this, truth be known, so that today's unsold tomato (unsold because it wasn't ripe) will survive long enough to be offered at another market tomorrow or the next day.

The farmer would not want to be caught bragging because he grows his own share of pretty bad tomatoes. Some of them have blossom-end rot because he gives them inadequate water or he is too blasé about the condition of his soil. Some of them are misshapen because he takes too many risks with the weather, setting the plants out too early.

And from time to time the farmer has fallen into lockstep with his friends. He succumbs to the claims of the seedsmen and he plants varieties that promise a world without end. He accepts the advice of friends and tries this or that variety and, invariably, he ends up hating himself.

So he goes back to basics, reverting to what he knows and likes. He grows mostly garden-type tomatoes, the kind that have been successful in backyards from coast to coast, and his rewards are great. His customers come back week after week, lining up for the tomatoes.

They are sometimes cat-faced and ugly. They are sometimes oddly shaped — at least not the picture-perfect beauties that line the pages of the catalogues. Often they don't reach the size the farmer wants and expects. But he picks them mostly ripe, and therein lies the difference.

One of the farmer's great pleasures is the man who comes early to the Sunday market, skulking around the back of the van, scanning the pile for the tomatoes that will make his week worthwhile. Last year, the man announced that these were the best he had ever eaten and he returned for more every week.

There was no question about the very first Early Girls of the 1989 season. They were automatically set aside for this man and handed to him at no charge as a token of appreciation, past and current.

Nowadays, the man shows up early each Sunday and simply passes the signal. "Pick me out some tomatoes that I'll like and hold them back," he says. "I'm going to get my breakfast and I'll be back when the market opens."

Now this is all well and good, but there is no room for smugness. On the advice of a valued friend, the farmer several years ago decided to grow a commercial tomato called Mountain Pride. Catalogues and growers gave it rave reviews.

The taste wasn't great, but it wasn't bad, either. It was prolific and it grew to a nice round shape. It was a tomato that seemed to work and the farmer, without reflection, was all set to grow it again this year.

Then rather full of himself, he ticked off his list of 1989 tomato varieties to plant breeder Nicklow. The Mountain Pride drew such a negative reaction from the man who eats thousands of fresh tomatoes every year that the farmer felt small and humiliated.

He took the hundreds of Mountain Pride seedlings that were burgeoning in his greenhouse, waiting for the proper planting size, and threw them on the compost heap. In their place went Whoppers and Jetstars and Early Girls and other tried-and-true varieties.

This, admittedly, was a harsh step, but Dr. Nicklow's lesson was valuable: Go with what you know, concentrate on taste, make your customers cry for more. The quest for the right tomato goes on, but in fact it's not that elusive.

BEAUTY IS ONLY SKIN DEEP

Jan. 25, 1989

A heavy snow, coming as a welcome surprise, has sheathed the fields in icy white and sent the farmer indoors to ponder a pot of black bean stew, pursue his mounting pile of seed catalogues, and weigh other matters of agro-culinary portent.

The radio brings news of massive traffic difficulties in the distant nation's capital, but out here it's just fine, thank you. The snow plows won't come by for hours and the isolation is delicious.

Which allows time to deal with the question that city people seem most wont to ask: "OK, what do you really do in the winter?" The tone usually implies, not without a certain envy, that for most farmers the cold months must be a sinful carnival of slothfulness.

The question has little to do with reality, unless the farmer is prosperous enough to chase the sun to southern climes. Work continues on the farm, albeit at a slower pace and of a different sort.

Animals still must be tended, equipment repaired, and ledger books updated. Planting schemes and finances must be arranged; seeds and soil nutrients must be ordered. A greenhouse cover shredded by a vicious wind must be replaced, a barn roof nailed back in place, a once majestic cherry, now toppled, sawed into firewood. The list is without end.

So a crippling snowstorm is welcome, providing a small respite, a time to think. To think about the pot of black beans, for example, which is not doing so well.

The problem is that the beans don't taste right, even though the farmer has not deviated from his traditional recipe. The insipidity is traced to the green peppers, which must be thrown into the pot in large quantities to achieve proper zest.

The peppers in this pot are supermarket peppers. They were bright and properly "blocky," as the plant breeders like to put it, and they were grown someplace far away for cold-weather cooks in Pennsylvania. They were purchased only because the farmer, imprudently, did not freeze his own when the time was right last summer.

But the bottom line is that these peppers have no discernible taste. In all probability they were bred with appearance, shelf life, and "shippability" foremost in mind. Moreover, they probably were swamped with irrigation water and heavily fertilized to make them big and get them to market quickly.

There is no great secret about the impotence of these peppers or, for that matter, about the wimpishness of dozens of other fruits and vegetables commonly found in the American grocery store. The supermarket tomato has become the great symbol and whipping boy, but in fact the Grand Bland is everywhere.

Authoritative panels of the National Academy of Sciences and the Department of Agriculture in the 1970s and, more recently, writers such as Jack Doyle of the Environmental Policy Institute have amply documented the problem.

It is simply that today's plant breeders work zealously to develop varieties that meet the needs of large-scale growers in the West, the South, and Mexico who produce the bulk of our fresh winter produce and large amounts of our processed varieties.

They seek to develop varieties that withstand the pummeling of machine harvesting, that resist more pests and diseases, that ripen uniformly, that have a long shelf life, and that retain their cosmetic appeal after cross-country trips in refrigerated trailers. Taste and nutritional content are the first elements to go when breeders start tinkering with a plant's genetic chain.

In their studies of the amazing transformation of the tomato into a plastic blob, Doyle and Jim Hightower, the Texas commissioner of agriculture, tell how geneticists sacrificed the tomato's vitamin content to achieve the other traits sought by growers and processors.

This year's commercial seed catalogues, now cascading about the kitchen table, provide more somber evidence of the trends. Growers are bombarded with dozens of new vegetable varieties for which soaring claims are made. But rarely do they allude to taste or nutrition.

Here, for example, is a new "profit-conscious" honeydew melon, touted for "its thicker wall, a profitable feature that enables the grower to command a higher price because of its extra suitability for shipping." Comes also a new Eastern cantaloupe "noted for its shipping and fruit holding qualities."

Now we have a tomato that is "firm and jointless for Florida and California." Another company offers a tomato designed to replace a variety put on the market only a few years ago, hardly time enough for growers to know if they like it or not.

New carrots are big in the books, but no one is promoting a carrot that above all else tastes like a carrot. "Heavy tonnage ... vigorous tops ... good interior color" are the key words. And we now have a new watermelon "bred specifically for boxing," which ought to please pugilists from coast to coast.

Not that all is gloom and doom, however.

One catalogue offers an improved type of Sweet 100 cherry tomato, already one of man's finest concoctions, which supposedly resists the cracking that makes it hard to market at farm stands.

Another offers a new squash that promises large golden male blossoms for the restaurant and kitchen trade. Some of the new sugar-enhanced sweet corn promises a two-week shelf life, a boon for consumers accustomed to corn turned to starch before they get it to the table.

So that, in part, is what farmers really do in the wintertime – study their catalogues and look for inventive new ways to grouse about things in general. At this small truck farm, before the snows are gone, there will be more grousing and long hours will be devoted to comparing the prices and claims for seeds, looking for varieties that will satisfy both grower and consumer.

Today it meant calling a seedsman in New York with questions about his catalogue claims for an apparently "new" Jetstar tomato, a favored old variety here. Nothing new about it at all, the man said; how many seeds do you want to order?

Yesterday it meant calling a potato grower in Maine, in search of the Yukon Gold, a tuber that is fast becoming a legend for its purported quality. Word was that Yukon Golds would be available, but to talk business we'd have to wait until the grower returned from Florida.

Grumble, grumble, grumble.

TRIAL AND ERROR BEATS
TRIED AND TRUE

S pring planting has begun with a fury and once again this is the time of tension in the truckpatch. The machinery keeps breaking down, the ground is either too wet or too dry, some needed seeds still haven't arrived, other seeds are crying to be planted, the daily work list is always too optimistic, the sun sets too soon.

Truth is, as the farmer has come to understand, the rest of the year, even well past the first frost of fall, is going to be this way. Too much to do, too much to keep track of, too much of everything but rest.

And yet it is a time of excitement, for a new season is under way and the farmer is buoyed by the hope of another abundant year and one more round of the serendipitous discoveries that make the adrenalin run.

The farmer surveys the scene, and for want of a better analogy, occasionally compares himself to the mad scientist who repairs to his little laboratory to work his alchemies and to seduce the plant divas that some far-out husbandmen write about.

It is not that the farmer has gone completely over the edge, however. It is that the longer he works with his land and his plants, the more he wants to know and the more he wants to plant and try.

Now, if he were a conventional grower on a conventional farm, he would plant his oats and his corn and his soybeans and be done with it, neat and clean. But it is not that way in the truckpatch.

Here, the key is diversity and a constant searching for new items that not only will do well in this climate and soil, but which also will appeal to the farmer's customers. This leads to excesses that often defy all reason.

The eggplant is a good example. In the beginning, the farmer grew what all his friends grew, a variety called Black Beauty. It was an eggplant that seemed to meet the basic requirements – reasonable taste, hearty, good-looking, not difficult to grow.

And then it occurred to the farmer, because the chefs and some of his customers told him so, that if one just looked, there was a better eggplant out there. The search began.

Another variety called Dusky soon joined the stable. It was so good that Black Beauty was abandoned. Then the farmer learned about the longer, slender Asian eggplants, noted for their taste and fine textures, and he turned to something called Orient Express. It was so good that it became a yearly regular.

But the discoveries kept piling on top of each other. The farmer's friend D'Amato brought from Italy the seeds of a Florentine eggplant, an oddly shaped lavender thing that had the creaminess of an elegant custard. That, too, became a regular.

Another friend raved about a miniature eggplant and the farmer felt compelled to try it. True enough, it grew well and it produced abundant little fruits. But it had a flaw — a bitterness that made it unappealing — and the hard decision was made to drop it.

More word-of-mouth enthusing and more ideas gleaned from the winter reading of seed catalogues have sent the farmer in new directions this spring. He has added another Oriental eggplant; he has seeded an obscure pink variety, and he is growing a Thai round and a Thai long eggplant, both preceded here by rave reviews.

Several thousands of these various seedlings — far more than the planting schedule envisioned — grow apace in the greenhouse and soon it will be time to take them to the field. The farmer's hunger for the different and the new has created a problem: the space allotted for eggplant is not nearly enough. Something else will be displaced to make room and once again, the farmer's carefully drawn rotational schemes will be disrupted.

Similar things occur each year with the tomato as the farmer keeps looking for the perfect fruit, the best tasting and the most easily grown. From the four or five varieties grown here in the beginning, the list now surpasses three dozen and the farmer has not an idea about where they will all be planted.

So it goes with other vegetables — peppers, greens, lettuces. The farmer keeps trying more and more of these things, looking for the ones that taste best and grow best. Each year he seems to try more of them than he abandons and his space problems grow exponentially.

In part, the farmer is driven by his desire to experiment. But in equal part he is trying to anticipate his market and his customers' tastes and their own quest for more health-giving vegetables.

In the beginning, because that was all the farmer knew, the standard curly southern mustard was grown and it did well. But there's a great range of greens out there, it turns out, and the truckpatch now is seeded to such things as Chinese swollen stem mustard, kailaan, tah tsai, San Pasquale dandelions, michihili, red Russian kale, shungiku and other unpronounceable items.

Of course, there's not enough room for all these things. Of course, there's not enough time to tend them all properly. And no, it's not entirely certain that all of them will sell enough to justify the effort.

But, the farmer keeps telling himself, you won't know unless you try. It may increase the tension levels, but it doesn't seem to be mad science at all.

WINTER HOLDS THE SEEDS
OF SPRING

Dec. 20, 1991

It happens almost without fail as the days begin to lengthen, even as the worst of winter is yet to come. The farmer starts to sense the rise of juices that presage a new year in the field, and much like an animal from hibernation, he begins to fidget and worry and betray a restlessness that means he must get back outside.

In reality, the winter break in the truckpatch is a misnomer. It doesn't begin until around Thanksgiving and it ends in late January, when the first seeds are sown in the greenhouse.

Occasionally, winter break means sleeping late of a morning or stealing an afternoon nap. But only occasionally. In truth, there's never time enough to decompress from the season just passed nor even – thankfully – time enough to dwell on its failures and disappointments.

Actually, the farmer has come to believe that he is the victim of a dark conspiracy orchestrated from the swank executive offices of some faraway seed company that begins bombarding him with temptations in November.

As the seed catalogues begin flooding the mailbox, the farmer's genetic alarm bells start ringing – it's time to order, it's time to move, time to till and plant. Of course it won't be time to till and plant for months, but the excitement brought by these provocative missives is enough to stir the somnolent beast.

By the time of ordering, the winter discussions with Peterson, the partner, have fairly well defined the new things that will be tried and the old things that will be dropped. She's babbling these days about sea kale and cardoon as exciting new ideas and the farmer is rolling his eyes ... but what does he know?

The experts at the agricultural schools caution farmers to never grow things for which they have no market. This may be true for corn and wheat, but it doesn't work for the specialty grower who tries to entice customers to his stand with the new, the different, the tasty.

As the farmer's friend Planck says, these people end up buying and trying things they never knew existed, let alone knew they needed. Thus the truckpatch will be expanding its growing line of oriental greens and salad items because customers, Asian or not, buy them with abandon.

So the catalogues that feature these items must be studied closely and the sales possibilities of each considered. And then comes the task of learning how to grow them and counseling how they are to be used in the kitchen.

And there are visual considerations. The chocolate bell pepper, the color of a Hershey bar, may be cute but its flavor lacks, so it was dropped from the mix long ago.

Other colored peppers – purple, orange, canary, lavender – are grown for their taste and the esthetic verve they add to a farm stand display (and ultimately, to a salad).

And there are simple bows to market demands. Takoma Park, for example, doesn't much care about head lettuce – probably a political statement – so the truckpatch's leafy offerings are just fine. But Hagerstown wants head lettuce and thus it will be added to the truck-patch seed order for next year.

Columbia may dote on a superbly sweet and smallish Japanese white turnip – grown here almost exclusively – while Hagerstown insists on the more familiar purple-topped variety. Purple tops will be grown again in the truckpatch.

Tomatoes pose greater challenges. The annual search through the catalogues sends the farmer ever deeper in to the arcanum of tasteful tomatoes. Several dozen varieties are tried each year; some are dropped and some are tried again.

But this search has demonstrated that the best-tasting tomatoes are the old-timers, the non-hybrid types that Granny grew but that are too disease-prone or too fragile to appeal to supermarket tastes. So each year, the truckpatch grows more of the old-timers.

The appearance of these zesty but sometimes odd-shaped varieties is enough to cause frenzies in Takoma Park and Columbia, with shop-pers elbowing each other away from the stand. In Hagerstown, they're asking for a more ordinary-looking tomato. So that, too, will be grown next year.

Not to misunderstand what the farmer is saying here. He grows what pleases his own palate and he hopes that those gustos will carry over to his customers. Years ago he couldn't give away sweet basil in Hagerstown, while metropolitan folk bought all that he could grow.

But he kept pushing his basil and, after a time, Hagerstown began asking for it. Nowadays, he can't give away his arugula there, but they fight over it in the metropolitan markets. So he'll continue to grow his arugula for the big city and something else for Hagerstown. Maybe just plain old head lettuce. Come the millennium, Hagerstown will beg for arugula.

That, the farmer must confess, is part of the fun of this strange business of growing and purveying vegetables. He finds enjoyment in promoting the offbeat and the unusual, which always must meet his own standard of good taste. He is hurt when his favorites are spurned, but he gets vast pleasure from the respect implicit in the customers' acceptance of his wares.

So now the catalogues pile up and the tempting challenges spill off the pages. City palates wait to be teased, tempted, and seduced. It's no wonder the juices are running again – the farmer, you see, relies on a supermarket for his winter produce as well.

A PLAN FOR ALL SEASONS, A CONTINUING HARVEST

Sept. 5, 1990

M ost of the vegetable gardens hereabouts, if not already plowed under, are on their last legs. But in the truckpatch it is almost as if spring has just begun.

Over there on the hillside, broccoli and cauliflower plants are looking like a million dollars. Down by the road, rows of mustard and turnips and radishes are poking up through the soil. Out back, beds of lettuce and other cool-weather crops threaten to flourish.

In other fields, the third and fourth planting of tomatoes have set fruit and give every indication of intending to produce before the first good frost takes them down for the count.

In the greenhouse, which the farmer doesn't even want to think about, there are seedlings of Shasta daisies, German statice, onions, rapini, lettuce, and cilantro that still must go to the field. And for good measure, 100 pounds of garlic awaits planting in early October.

The farmer's city friends and acquaintances, accustomed to the simple cycles of the home garden, often talk about "the harvest" in the truckpatch as if it were a fixed event occurring on schedule each year.

"Which harvest?" the farmer is wont to reply, for the harvest actually goes on constantly from April until the end of November. With dozens of kinds of vegetables, fruits, and flowers going into the ground over the course of a season, something is always ready for harvest.

That continuity of supply, of course, is the bread and butter of the truckpatch. Making it all work, more or less on schedule, is the hard part, for the weather and the clock rarely cooperate. In her playful way, Nature makes a frustrating game of this.

Now, as fall approaches, the farmer smiles at the memory of innings played and won by his side. He remembers sensing the oncoming summer rainstorms and rushing to a field – dropping the task at hand – to put more beans or whatever in the ground to take advantage of the new moisture. One year the beans came up in four days and there was no way to measure the farmer's smugness.

He now takes pride in the broccoli crop that burgeons on the hillside, but he has to smile about that, too. Memory tells him that he groused and muttered when Peterson, his partner in grime, insisted the broccoli absolutely had to be planted in the driving rain of a July day. She was right, as the fall crop keeps telling him day after day.

And yet, great bursts of excitement spring from this game. The sight of a germinating seed struggling to push up through the soil, be it a simple radish or a persnickety carrot, never fails to thrill the farmer and bring him satisfaction.

In a way it signals that he has done something right – put the seed at the right depth, picked at the right time, chosen the right variety, taken all the proper care and precaution. It is humbling as well, for the farmer is awed by the power of a seed and knows that he is not much more than a bit player in the drama of its growth.

If the harvest is a constant thing, so is the planting. Looking back at his journal the other day, the farmer was astounded at what he read and calculated. The first seeding, albeit in the greenhouse, was on Jan. 18; the last, which will be lettuce, will occur next week (mid-September).

Before the year is ended the farmer and his accomplices will have put roughly 80,000 transplants into the ground – every one by hand. Each one of these was started in the greenhouse by carefully dropping into a plastic cell a seed, which was then meticulously watered and fertilized.

This does not just happen. Back in January, Peterson and the farmer talked at length about their plan for the year and then devised a sophisticated, possibly unrealistic, schedule for greenhouse seedings and outdoor planting. All of this assumed an abundance of time and perfect cooperation from Nature.

The schedule, for example, called for 14 greenhouse seedings of lettuce. This, the farmers calculated, would provide a steady supply through the year and maybe, for once, allow them to have lettuce in the impossibly hot weeks of summer.

In the cool of early spring, things went swimmingly. February-seeded lettuce went to the field on March 10 and was sold at market on April 22. The farmers exulted in their triumph.

The plan looked good on paper, but the reality was something else. Two seedings were missed because there simply wasn't time. Intense heat fried several more plantings, even though they were kept watered and covered with long green shade cloths. And what the heat didn't get, the marauding deer did.

There were other frustrations. The first planting of beans in April was a blank – too much moisture spoiled the seeds – and threw the year's entire bean-planting schedule off course. And now, the final planting of beans looks world-class and the harvest crew has gone back to school or somewhere else. The farmer will pick the beans.

March-planted radishes failed completely (too wet, too hot, too many flea beetles). July-planted carrots did nothing (too dry). The radicchio planted in the spring bolted from the heat before it could grow to harvest size. Ditto for the fennel. Rutabagas planted too late (the ground was too wet) failed utterly. A couple of plantings of beets made it; a couple more did not.

The coolness of May slowed the early tomatoes and then a blight induced by damp and cool threw them into a tizzy. The second planting grew phenomenally in a low-lying moist patch, just as the farmer envisioned, but ripened late because of the shade from a practically impassible mass of vines.

So it goes in the truckpatch. The frustrations of failed plantings are felt nearly every day, but they are assuaged by successes that seem to outweigh them. And they are tempered by the thrill that still comes from seeing a seed burst into life.

That thrill is seductive, tempting the farmer to take chances that make no sense. Rummaging through his seed supply in late August, he came upon a stock of beans he didn't know he had. There's still time, he rationalized. Plant these guys, he told himself. Get a crop.

And then reality grabbed him by the neck. Of course, they would germinate. And of course, they would get frosted out. Or they would not get picked. The farmer put the beans back in the can and tightly, definitively, jammed the lid into place.

The thrill isn't gone. It's just that enough is enough.

THE FIELD OF DREAMS

Oct. 11, 1989

It came to be a ritual joke: As summer turned to fall and the rush of the harvest intensified, the farmer would tell friends and customers how he was praying for the frost that would deliver him from his burdens.

They would laugh at this line, for no one in his right mind really wanted to see an end to the abundance from the truckpatch. But in his heart of hearts, the farmer was only half joking. He craved a rest.

From early March into the cooling mornings of October the season had been running at full tilt. With the exception of an odd day here or there, the farmer had not paused. He could not pause, for the work seemed endlessly demanding. In a sense he had become a slave to the farm – a willing subject, to be sure, but still a slave.

So the farmer was weary now and the idea of a killing frost – once an event that hurt and disappointed – became alluring and warming. The soil would go to rest for the winter and the farmer would wallow in long days of lassitude that would bring the pleasures of sleeping late, reading books, and napping at will.

Most of this is fantasy, for the farmer's year is far from ended. The cool of early fall, especially after the first touches of frost, is exactly what some of the crops here want most. The broccoli, the cauliflower, the greens, and the lettuces keep right on going. The root crops go on until the ground freezes hard.

As these crops are brought from the fields, the pleasureless chore of putting the farm to bed for the winter begins. This is the hardest part. If late October and November behave as can be expected, it means the farmer will be toiling in cold, rain, mud, and maybe snow and ice. It means perpetually wet feet, perpetually numb fingers, biting wind in the face.

Weather in the early spring often is identical, but at least the farmer is energized by the prospect of a great new year, thrilled by the signs of nature reasserting itself, seduced by the pungent aroma of freshly tilled soil and the first sighting of bluebirds and robins.

Now the farmer is driven mostly by the thought of the midday nap. But before he can join his fields and the resident groundhogs in hibernation, he must grapple with the end-of-season drudgery.

For example, the farmer this year used 29 rolls of black plastic mulch. The farmer could not operate without the plastic, but he despises it. The plastic is not reusable, so it must be removed from the beds and disposed of by special arrangement in a faraway dump since there is no public landfill in the farmer's county. And since the farmer has no machine for removing the mulch, he must pull it by hand – a frustrating and usually muddy, filthy job. The long strips tear, shreds of plastic flutter about in the wind and must be plucked from the fence rows where they collect.

Twenty-nine rolls of black plastic mulch translates to about 11 miles, about the distance from, say, Annandale to the U.S. Capitol. This is a lot of plastic to be pulled by hand. Just the thought of it makes the farmer want to take a nap.

Once the black plastic mulch is removed, the farmer then must gather up the plastic drip irrigation tapes – about six miles of them – that have watered his beds and protected against drought. Many farmers simply throw the tapes away and buy more. But because this farmer is frugal, he will save and reuse as much of the tape as he can. Again, it means mud, filth, and frustration, for some of the tapes will break and blow in the wind.

The farmer will roll out the cumbersome cable spools he got from the telephone company and try to reel in the big hoses that carry water to his drip tapes. He will inevitably tear the hoses in a few places or break their plastic components and he will mutter unprintable things.

When the growing beds are cleared, the farmer will determine which of them require treatment with fertilizers. He will sprinkle rock phosphate and sea kelp by hand from 50-pound bags (yet again, he has no machine for this) and he will apply his composted chicken manure to give the beds a nutritional boost. This, too, will be done by hand, for his manure spreader flings out the stuff too liberally.

By then, no doubt, the daily chill factor will have increased and outside work will be even less inviting. But the farmer will mount his tractor, squint into the cold wind, and haul his tilling machine out to the beds to turn the soil and incorporate plant residues that will enrich its organic content.

To prevent winter wind and water erosion on the beds that crisscross the hilly farm, the farmer then will broadcast by hand wheat and rye seeds that will germinate before the big freeze and create a protective green blanket. In the spring, this residue will be tilled back into the beds to add to the tilth.

And then there is the new ground that must be considered. The farmer will cut new beds in rested sod that will go into production next season. Each of these new strips will get the same treatment of soil amendments and seeding to create a winter cover.

Along the way, the farmer will pull up and store the hundreds of stakes that held up his tomatoes. He will pull, roll, and store about a fourth of a mile of trellises that supported a fall sugar snap pea crop that didn't grow for squat. He will clean out the greenhouse and sterilize the hundreds of plant trays that held his seedlings.

If there is time and energy – and there must be, for these are priorities – the farmer will gather and store the vegetable boxes that litter the packing shed and obstruct passage in the barn and garage. He will inventory his packing supplies and begin to think about his needs for the new year.

The farmer likes to fantasize that he will move a little heater into the barn and tinker with his machinery, tuning it for spring and painting out the rust. He will sharpen and oil his hoes and scythes and other weeding tools. He will, he thinks, finally clean up his work bench and locate lost wrenches, screwdrivers, and pliers.

All of this awaits the farmer after the killer frost that he joked about, even yearned for, in September. There is a certain amount of insanity about the whole business, for no matter the yearning, no matter the intentions, no matter the weariness, the work won't be done until the farm is finally put to bed for the winter.

So more likely than not, the farmer's midday naps will be deferred until the deep cold of January. But then the catalogues will be arriving, seed orders will be pondered, and the promise of a new season will become the driving force ... which is another story.

THE FARMER PLOTS
HIS ANNUAL GAMBIT

March 21, 1990

The unseasonably warm weather came suddenly, timed perfectly to quench the farmer's winter-born thirst to get to his fields, and the first of the lettuce transplants and onion sets went into the ground without a hitch.

The early onions went to a plot that last year grew hot peppers, the lettuce to plots that held cantaloupes, but it was then that the great annual dilemma began.

The dilemma in the truckpatch is knowing where to plant things. A basic rule of farming, although ignored with dismaying frequency by many growers, is rotation of the crops.

By assuring that the same crop is not planted in the same place in consecutive years, the prudent farmer learns that insect damage is reduced and that his soils are less depleted. And by resting a portion of his fields each year, his land can be replenished and his weed problems diminished.

That is not so difficult for the grower who produces just a few things – corn, soybeans and small grains, for example. There's less to keep track of and if he happens to be a chemical farmer, he may eschew rotation completely and rely on pesticides, herbicides, and synthetic fertilizers to carry the crop.

But with more than 100 different vegetables, fruits, flowers, and herbs to tend in his truckpatch, the organic farmer who does not use the chemicals confronts a daunting and often maddening challenge of superintendency.

He knows that the Solanaceae family – potatoes, tomatoes, eggplants, peppers – must have a different home each year. He knows that the brassicas – cabbage, broccoli, and cauliflower – must not follow brassicas. And he cannot use the same field two years running for the cucurbits such as melons, squash, cucumbers, and pumpkins.

Moreover, each of these crops would like to be planted in a field previously given to legumes – beans and peas, clover and alfalfa –

because of the valuable fertilizing nitrogen they pluck from the air and fix in the soil.

The problem, however, is that the farmer just doesn't grow enough legumes on his vegetable land to make this rotational mix possible. So the farmer must improvise and juggle to make the best use of his land and to give each of his many crop varieties a fighting chance.

The picture becomes additionally complicated by the plant's growing habits and their nutritional needs. Some want acid soil. Some want sweet soil. Some want more nutrients than others, some want more sun. Some, such as the early and late greens, are magnets for marauding deer and groundhogs and must be planted as far from wildlife runs as possible.

Still others cannot survive without regular watering, so must be assigned to plots that are assured of irrigation.

The farmer tries to keep track of all these elements with elaborate lists that he compiles on yellow legal pads during the winter. But in practice, when planting time arrives as it has now, reality always sends the farmer back to the drawing board. This year's tomato plans are an example. Ideally, the tomato would go to fields with ample water and moderate fertility, but no matter how the master plan is revamped, the farmer simply cannot find a way to reach that ideal for each of his three plantings.

The first planting will go to an irrigated field that contained flowers in 1989. The plot's fertility has been deemed adequate for the tomatoes, but to protect against blossom-end rot in the fruit, the farmer knows he will have to add lime to the soil. So far, so good.

But the second planting of tomatoes, which is certain to mature in the deep heat of summer when water is scarcest, will have to go to a low-lying unirrigated field that contained melons last year. In a drought, it would have been ideal for melons because it lies in a moisture-attracting hollow. But too much rain was the problem last year and the melons were blitzed with fungal diseases.

So the tomatoes will be planted with a heavy straw mulch to help retain moisture and the farmer will keep his fingers crossed that too much rain does not come at the wrong time. But just in case it does, the farmer also knows that those 1,800 plants will have to be individually tied to stakes to help avert early blight or later damage from the wet.

The third planting of tomatoes will go someplace else. The farmer isn't sure where, because his scrupulously drawn plan on the yellow pad has somehow failed to allow a space.

The brassicas pose an altogether different problem. The very best spot on the farm for early broccoli transplants, which should go to the field next week, is a hillside that absorbs copious amounts of morning sun. The soil seems just right, but alas, it contained purple cauliflower last year and thus cannot have broccoli.

Last year's fall and spring broccoli was planted in irrigated fields that have been assigned to peppers and lesser crops such as leeks and beets this time. The farmers' only solution for early broccoli is a field without water out near the woods, which harbor a growing population of deer.

This is another gamble, the farmer knows. But it's the only way a rotation will work. A blanket of mulch may keep the broccoli moist enough to produce a nice June harvest. But there is no way, short of keeping a howitzer at the ready, to hold the deer at bay.

The lesson was taught painfully in 1989 when the farmer, desperate for a place to put his fall lettuce and radicchio, settled on a field not far from the woods. The plants did beautifully, but then, as the woods dried and other greenery faded, the deer marched down the rows and chomped off half the crop.

So the lettuce and the radicchio this year will be kept far from the perimeter, hopefully hidden well enough among other plots closer to the outbuildings that the deer will dare not venture near. Nice plan, but other greens thought to be well hidden this way last year also were systematically mowed down by the predators.

Imagining the truckpatch as a giant game board, increasingly crowded with players and pieces, is not far afield. But the dilemma of where to plant things and remain faithful to the rotation is not the only one facing the farmer. Early blooms of the forsythia and the daffodils make the farmer nervous.

Temperatures better suited for May than for March fuel the urge to get as much in the ground as possible. The old-time gardeners assure that the first leafing of the lilac means it is time for the onion sets and the peas to be planted.

The lilacs have popped forth, and the onions in fact were duly put in the soil, but the farmer knows he must restrain himself. He knows in his bones that winter is not gone and he must curb his zest to sow the early crops. But then, what if winter really is over? What if an early spring is not taken advantage of?

The dilemmas seem never ending.

The farmer's friend Pollock, who raises vegetables near Allentown, visited recently with an impressive sheaf of tables and charts that he offered as a way out of the quandaries.

He had found a way to computerize his rotational scheme so that each of his growing areas was assured of a rejuvenating rest every three years.

Each growing area was assigned a number and a letter and by moving these around annually, Pollock was certain he had found the perfect way out of his own dilemma. No more yellow pads, no more panic at planting time.

On paper, it seemed to be an excellent approach. The farmer pored over the charts, studied the tables, and weighed the appeal of bringing order to his life. He finally set the charts and tables aside and turned back to the yellow pads, knowing that in this truckpatch, at least, things are just too complicated for a computer to handle.

A Midwinter's Daydream

Jan. 3, 1990

The deep freeze has driven the farmer indoors to warmth beside the stove, and thoughts, however hallucinatory, turn toward an imaginary future of Junes in January when plants might flourish in the cold.

Each year, increasingly, the catalogues offer new items, including curiously named greens from the Orient that are hailed for their nutrition and their resistance to cold. Some thrive under a blanket of insulating snow; some endure the harshest blasts of winter. This sets the farmer's mind to wandering and wondering.

His thoughts are provoked by the snow cover on the fields. In his mind's eye now he is seeing healthy, hardy green plants poking up through the white. And as he daydreams, he hears the faint rumble of the eastbound semi-trailer trucks on the distant interstate hauling fresh vegetables to city salad bars.

The big trucks seem to roll along endlessly. They are headed to markets usurped by industrious farmers in the country's warmer climates – farmers who have changed our eating habits to the point where the produce of June is always available in our Januaries.

Well, the farmer is thinking, perhaps maniacally, that it needn't be this way. Someday, he posits, that abundance of water in the agricultural West may run out. Someday, another energy crisis could slow or stop the semi-trailer trucks.

Someday, the thought continues, we might be growing salad greens in the dead of winter in the East. And necessity and demand might give rise to a whole new generation of small-scale family farms doing what they did decades ago – servicing their local markets and feeding the people nearest them.

Unless one subscribes to all the global warming theories, this might seem to be just so much poppycock. But in fact it is not all that far-fetched. Time was when we got along fine without greens from the Salinas Valley and winter cantaloupes from Arizona.

So, the farmer asks himself, why aren't we turning back the clock? The 19th-century horticultural books that are the basis for much of the farmer's learning and procedures tell about a day when America fed itself year-round without imports from the warmer parts of the country or Mexico or the Southern Hemisphere.

In that era before refrigerated trucks and trains, the large Eastern cities were ringed by small truck farms, which used hotbeds, glass-covered forcing houses, sash houses, and ingenious heating systems to produce a bounty that satisfied consumers' palates.

The books tell of similar systems in Europe, particularly in the environs of Paris, where farmers routinely maintained year-long crop production by using protective covers to extend their seasons.

Peter Henderson, a distinguished New Jersey market gardener, wrote in the 1860s about fall planting of lettuce, cabbage, and cauliflower under protective covers to get crops well before winter's snows had ended in the East.

Henderson also reported great success in seeding more tender crops such as tomatoes and peppers in the dead of winter – always protected, of course, but destined to be in before the traditional plantings.

Consumers accustomed to going without these products in the cold months paid accordingly for the new bounty. Henderson reported that the profit, despite extra production costs, was inducement enough for farmers to challenge the elements.

Henderson wrote that the potential profit from winter crops was so great that "there is no doubt that in hundreds of cities and towns of the Union the same use of sashes would double or treble these results."

Today, not many Eastern growers would even consider this labor-intensive, off-season production, unless they were working under cover of expensive modern greenhouses.

But inspired by Henderson's example, the farmer tried an experiment last year. On a warmish day in January, when the soil was frigid yet workable, the farmer seeded a large bed of lettuce. Friends and neighbors chortled and chided the farmer for his obsessiveness.

The farmer was not deterred. He covered the bed with a blanket of polyester fabric, to hold in moisture and give some protection from the wintry blasts that surely were to come, and then moved on to other preoccupations.

By the end of February, tiny lettuce seedlings were pushing up through the soil. By March, the plants were beginning to flourish and by the end of April, a bounteous crop was about ready to harvest.

If it can be done on this small scale, why not en masse? With the multitude of plastics and plant protectors now available to agriculture, why were Eastern farmers not taking these risks?

The answer, in all likelihood, is that most farmers are either too fatigued from the rigors of the traditional growing season or too intimidated by competition from the warmer climes.

But it could be done and there might come a time when it would have to be done. Hallucinatory, perhaps, but that's one of the things winter is good for on the farm, with the time it brings for conjuring and speculating about the future.

The rigors of cabin fever have afflicted the farmer's friend Planck, a perspicacious vegetable grower in Virginia, with even more terminal mental exercises. He foresees a not-too-distant day when the West, because of problems of water and environment, actually might be importing food from the East.

Peter Henderson would not laugh at this. For the nonce, however, at least in this winter of bitter cold, it all seems too far away, too visionary. More appropriate to pull a little closer to the stove and keep searching for a nice summer lettuce.

WITNESS TO A REVOLUTION

Oct. 31, 1990

The end of another growing season draws nigh and the farmer's mind, given time to wander during the long and tedious drives to city markets, is crowded with excitement about tomorrow.

In his mind's eye, the farmer envisions a new day coming for agriculture and food consumers in the East. He can see regional economies being boosted, rural towns rusticating, and farm families working together again in a partial return to the way it used to be.

The farmer's mind conjures more small farms producing more fresh fruit and vegetables for more people in the great megalopolis. He sees a day when there will be opportunity – and a need – for policies aimed at increasing the farm population to help meet local demand.

This is not so visionary, for in fact the revolution already is occurring in community after community in the region. The throngs of customers who attend the farmers' markets in the metropolitan Washington area, in places like Arlington and Takoma Park and Fairfax, bespeak the interest and demand for locally produced fresh food at prices that can be well below those of the supermarkets.

Many communities and neighborhoods indicate a desire to establish these markets, but the sad truth is that there are not enough growers of fresh produce to meet the needs. The trick lies in persuading more state and federal officials and agricultural academicians to become supportive revolutionaries.

A number of states – Massachusetts, Texas, Tennessee, and California, to name four – already are well ahead of the pack in creating dozens of state-certified markets, where local farmers are provided places to sell their products.

But there are signs that officials in Maryland and Virginia are beginning to catch the drift. The best example is Maryland, where on the heels of the creation of a successful market in Columbia this year, the state agriculture department has decided to set up five more markets in 1991. Virginia officials are being pressured to expand the "HomeGrown" concept that has seen the birth of 11 producer-only farmers' markets in Northern Virginia since 1980.

One of the obstacles to growth, however, is that no one, not even the bean counters at the U.S. Department of Agriculture, knows or has attempted to find out how much economic activity is stimulated by the producer-only markets. The answers are probably astounding.

The farmer's friend and guru, Chip Planck of Loudon County, a political scientist turned vegetable farmer, has been lobbying Richmond to learn more about this as a foundation for expanding the network of markets around the state.

Planck's tailgate math suggests that the Northern Virginia markets may be generating as much as $2 million annually in agricultural sales that otherwise would not occur without these community-supported outlets. He is urging a formal state study to determine the facts.

In case after case, as Planck notes, the existence of markets puts regional farming possibilities in a new context. The markets expand the season of local production, they increase the variety of things that are grown or attempted by farmers, they increase the types of people and operations involved in agriculture.

Looking around a market such as, say, Takoma Park, the farmer sees a carpenter and a construction worker, farming part-time, leaning toward becoming full-time farmers; a naval weapons expert selling honey from his bee and sheep operation; two journalists turned into successful bakers; a college-trained anthropologist making it with flowers and vegetables; a television repairman now an expert full-time grower.

They are able to exist and thrive because of the market. Too small to meet a supermarket chain's volume demands, they do just fine in a community outlet.

On their stands the farmer sees the usual array of tomatoes, melons, and sweet corn – all the traditional standards – as well as things that used to come from someplace else: Six kinds of hot peppers, purple cauliflower, rapini, daikon radishes, heirloom tomatoes and squash, Japanese turnips, Chinese mustard, French sorrel, kohlrabi, Italian green beans, miniature eggplant. And on and on.

These things all sell and they sell well, in part because of consumer interest in trying something different, in part because of the quality. And it seems, the greater the variety in these markets, the greater the traffic flow.

But the key, as Planck and other visionaries keep insisting, is that the markets succeed and farming opportunities increase because of strict producer-only rules that are in place. That is, if you don't grow it, you can't sell it. As long as hucksters can import products not normally grown in the region or resell someone else's produce, farmer incentive to diversify and compete is stifled.

There is another important, less tangible side to all this, as the farmer has sensed in his trips to the new market at Columbia. Urban customers understand that their patronage contributes to the local economy and that it stimulates regional farming activity. They are enthusiastic about the presence of farmers in their village and they feel a special sort of bonding.

That is why, as the farmer on Thursday makes his final trip of the year to Columbia, he will be feeling a certain sadness. It will be farewell to all these new friends whose names he does not know, but who have made him feel useful in their lives with his produce.

That is also why the farmer is excited about the more figurative tomorrows. As he makes that long trip to Columbia, he likely will be passed by dozens of refrigerated semi-trailers hauling lettuce from the West, as happens every Thursday.

The farmer knows, and many of his customers know, that it can't last and probably shouldn't. If nothing else, the high cost of diesel fuel makes long-distance produce cartage unwise and wasteful. It could be the best thing to happen yet to local agriculture.

EPILOGUE

How far, it seems, the official reckoning of small farms, local produce, and organic produce has come since the Flickerville Mountain Farm's beginning in 1983. Where once there were a handful of farmers' markets in the Washington metropolitan area, there are now several dozen – one even at the U.S. Department of Agriculture.

Supermarkets clamor for local produce, or at least produce that they can hawk as local with a modicum of legitimacy. An unthinking bureaucracy attempts to undermine the definition of "organic" in setting federal standards, and the public rises up in outrage, filing more than 250,000 comments – a record for public comments on a proposed federal regulation.

In 1998, the U.S. Department of Agriculture appointed a blue-ribbon commission to examine the plight and the potential future of small-scale agriculture. That might be seen as a small step toward the revolution that Ward dreamed of, although he would not have put much faith in the government leading the way.

All these events underscore his firm belief that a large and growing segment of the public cares deeply about the quality and safety of its food. Every year that he toiled over the plants in the truckpatch, sharing the experience with chefs and market customers, with rural neighbors and urban apprentices, he grew more firmly convinced of that.

The greatest difficulty, as he saw it, was that there were not enough farmers to satisfy the demand for fresh, seasonal produce. The truckpatch struggled each year to grow enough to keep pace with its growing clientele of restaurants and farmers' market customers. Invitations to attend new markets were tempting, but to expand into new markets would have meant sacrifices in either variety or quality, and that Ward would not do.

The obvious answer, now as then, is more farmers.

In recent years, there has been a swelling of public concern about the loss of farmland as farmers retire, die, or give up the financial struggle and sell their land to developers. Some states and counties have launched programs to buy development rights to farmland, hoping to preserve a base for agricultural production as well as open space.

There has been less concern about preserving farmers, the assumption being that new farmers will spring forth, fully capable, if they are needed in greater numbers. This is short-sighted.

Farming involves more than having access to land and the use of a tractor. Good farming requires knowledge, yes, but it also requires an ethical framework that can only be gained through a long association with the earth and all the creatures, human and otherwise, that rely upon it.

Farming used to involve a partnership between generations – the old teaching the young, so that when the time came for stewardship of the land to be passed on, the new stewards were ready. These days, fewer children of farmers are following their parents into farming.

There is a new partner at the Flickerville Mountain Farm & Groundhog Ranch – Brian Cramer, the bright apprentice that Ward wrote about in a 1992 Truckpatch column. It is my hope that when my seasons of farming are done, Brian will choose to take the farm onward.

What little I can teach him has already been taught in the last four years. His lessons now come, as Ward's did, from the land itself.

🐗🐗🐗

A few weeks before Ward's death, he attended one of those inevitable winter farming conferences, where he helped present a workshop for aspiring farmers. Ward enjoyed sharing his experience with "wannabes," and this time he shared the podium with Chip Planck of Loudon County, Virginia, who Ward often cited as one who had inspired him to the plow.

"I know how to get into farming," Chip said in his opening remarks. "I wish somebody could tell me how to get out of it."

The joke got a knowing chuckle from Ward. Difficult as it is to become a farmer – to accumulate the land, the equipment, and the myriad skills needed – it is harder yet to stop being one.

In the truckpatch, farming is a continuum. No annual event signals the beginning of a season, no final task marks its end. There is no logical stopping point.

Even winter is not a hard break, for there are plants in the ground and in the greenhouse. In the dark depths of January, there is the smell of warm earth and tiny green seedlings. There are winter "pet" days, as Ward liked to call them, when southerly breezes bring the feel of April to February.

Winter gives way reluctantly to spring, spring flows eagerly into summer. The farmer can only struggle to keep up with the segue of the seasons.

The beginning farmer circles dates on the calendar – reminders that peas and potatoes should be planted, fruit trees should be pruned, tomatoes set out. The experienced farmer doesn't need a calendar. His reminders are all around him, in the swell of a lilac bud or the cry of a nesting killdeer.

The lilac buds were swelling when Ward died on Feb. 23, 1995, two weeks after being diagnosed with pancreatic cancer. There was no question that the truckpatch would continue; nature had already deemed it so.

Ward Sinclair and Cass Peterson at farmers' market in Takoma Park, Md.

Other "Must Read" Books From American Botanist

THE GARDENING BOOK OF JAMES L. HUNTER,
A SOUTHERN PLANTER
By Catherine Howett

Planter Hunter's father was a general with Andrew Jackson. From antebellum Georgia, Hunter's 16-page garden diary from the year 1846 is complete with a numbered map that clearly shows what he planted and where he planted it in his 10,000-square-foot garden.

46 pages, paper.   **$18.00**

18TH CENTURY CLASSICS
WILLIAMSBURG'S JOSEPH PRENTIS:
HIS MONTHLY KALENDER & GARDEN BOOK

Describes planting and cultivating instructions for an American garden from 1775, including dates of planting and varieties that Judge Prentis sowed in his garden in historic Williamsburg, Va.

65 pages, paper.   **$18.00**

The ★★★★★★★★
AMERICAN
BOTANIST
booksellers

To Order, send check or money order to:

The American Botanist
PO Box 532, Chillicothe, IL 61523

Please include $4.00 for postage and handling for the first copy,
plus $1.00 for each additional copy. Illinois residents MUST add sales tax of 6.5%.

• **Phone:** (309) 274-5254 (10 a.m. to 10 p.m. CST) • **e-mail:** agbook@mtco.com
• **For fastest service,** FAX your credit card order to (309) 274-6143.

Charge $_____._____ to my credit card: ❏ VISA ❏ MasterCard
Credit card No.:_____Exp. Date:_____
Cardholder Signature:_____

Please rush me the following titles:
1._____
2._____
3._____
4._____

Visit us on the web at: www.amerbot.com